MONSIGNOR WILLIAM BARRY MEMORIAL LIBRARY
BARRY UNIVERSITY

0 2211 0022

D0914009

D
16.8   213173
.K38
1991

Msgr. Wm. Barry Memorial Library
Barry University
Miami, FL 33161

KEMP

ESTRANGEMENT OF T...

The Estrangement of the Past

THE ESTRANGEMENT OF THE PAST

A Study in the Origins
of Modern Historical
Consciousness

ANTHONY KEMP

New York   Oxford
OXFORD UNIVERSITY PRESS
1991

Barry University Library
Miami, FL 33161

Oxford University Press

Oxford New York Toronto
Delhi Bombay Calcutta Madras Karachi
Petaling Jaya Singapore Hong Kong Tokyo
Nairobi Dar es Salaam Cape Town
Melbourne Auckland

and associated companies in
Berlin Ibadan

Copyright © 1991 by Oxford University Press, Inc.

Published by Oxford University Press, Inc.
200 Madison Avenue, New York, NY 10016

Oxford is a registered trademark of Oxford University Press

All rights reserved. No part of this publication may be reproduced,
stored in a retrieval system, or transmitted, in any form or by any means,
electronic, mechanical, photocopying, recording, or otherwise,
without prior permission of Oxford University Press.

Library of Congress Cataloging-in-Publication Data
Kemp, Anthony, 1956–
The estrangement of the past :
a study in the origins of modern historical consciousness /
Anthony Kemp.
p. cm. Includes bibliographical references.
ISBN 0-19-506545-X
1. History—Philosophy. 2. History—Sources.
3. Historiography—Sources.
I. Title.
D16.8.K38 1991 901—dc20 90-6846

9 8 7 6 5 4 3 2 1

Printed in the United States of America
on acid free paper

D
16.8
.K38
1991

2/3/73

Preface

In 1648 or a little earlier, Thomas Hooker, minister and "virtual dictator"[1] of the Puritan church and colony at Hartford, Connecticut, wrote:

> Truth is the Daughter of time, was the saying of old, and our daily experience gives in evidence and proof hereof, to every mans ordinary observation. Only as in other births, so here, the barrennesse and fruitfullnesse of severall ages, depend meerly upon Gods good pleasure, who opens and shuts the womb of truth from bearing, as he sees fit, according to the counsell of his own will.
>
> Not that there is any change in the truth, but the alteration grows, according to mens apprehensions, to whom it is more or less discovered, according to Gods most just judgement, and their own deservings.[2]

"Truth is the Daughter of time"—"the saying of old" is an import from an other and an older world, England and a pre-Reformation past, and Hooker's reading of it is clearly a misreading. What it must have meant to those who repeated it to each other in that older world is, "Only that is true which has the authority of the past, of establishment by long ages of tradition." The saying is an argument against the new, against raw and untried novelty. Hooker's interpretation turns the old folk wisdom into its opposite: a fable of barrenness and parturition, in which there can be no eternal generation, but only an opening of the womb conditioned by a precedent closing of it. That "Truth is the Daughter of time" means to Hooker not that truth is contained in a wise past that extends continuously into the present, but that it is a sudden birth after an age of darkness. The ages of his past are "severall," not only multiple, but distinguished and severed from each other, periods of contrasted "barrennesse and fruitfullnesse." His is an age of "fruitfullnesse," the immediate past one of "barrennesse." The authors (if authors there can be of such communal epigrams)[3] of the saying may have felt an absolute intellectual identification with the past; Hooker can feel none: that is the meaning of his reflection upon the inherited text. It must be reread, reconceived into an implied identification with the Old Testament type of Sarah's womb.

"Not that there is any change in the truth," Hooker hastens to add—the almost sprung rhythm of the monosyllabic words forcing a haste in the reading voice—because the "alterations" in "mens apprehensions" certainly give an impression of a change in the truth. This implication must be gainsaid by a division between perceptions and the truth, according to deserts, and this makes the inhabitants of the past all the more damnable, willful dwellers in darkness.

Between the fourteenth century and the eighteenth, the Western comprehension of historical time reversed itself, from an image of syncretic unity and an essential sameness of time, to one of dynamic and supersessive change spawning schism after schism from the inherited text of the meaning of the past. Hooker stands at the fulcrum of the movement, an interpreter from the new paradigm handling a text inherited from the old.

The handling of texts is at the center of the change, for the notion of historical "truth" or "fact" cannot be understood apart from "mens apprehensions" of it. The past is by its nature removed utterly from all empirical instruments, and history can be no more than conceptions recorded in an immense palimpsest of historical texts: literary inventions, reinterpretations, attempted erasures. The past cannot exist as an object apart from the consciousness of it; changes in the consciousness of historical time are the work of literary inventions and critical interpretations of the preexistent tradition, and history is a fundamental literary genre, which tends to be denied its place as such because of its importance to ideology and its hold on the social and political mind.

This problematic juncture/disjuncture between history as text, narrative, writing, representation, and history as empirical science, has fractured the scientific historical enterprise inherited from the nineteenth century. Michel de Certeau, Hayden White, Paul Ricœur are inheritors of the historical problematizing of Rickert and Collingwood,[4] of the insistence that history is not a natural science, that the past presents nothing to our senses but a void which the imagination attempts to fill with webs of narrative. These insisters on narrative, on history as reconstitutive imaginative act, have been attacked as skeptics and historical relativists by such defenders of historical objectivity as Mandelbaum and Goldstein.[5] Between these poles ranges a debate, and a history of debate, of tremendous complexity. Even the defenders of a historical past independent of the accounts of that past problematize and attenuate its epistemic accessibility. Nor can the narrativists be entirely identified with skepticism and relativism. Augustine is a historical skeptic in the terms of Mandelbaum's accusation, and Ricœur derives his own insistence on the narrativity of history from Augustine. Both the essentialist and the existentialist tend to skepticism about the past-in-itself; the pale souls in between insist on its reality.

I do not intend to contribute to this debate, nor even to situate myself within it. I will assume, *ab initio,* that the past has no perceivable existence beyond its literary expression. My purpose is to describe the narratives and the narrative of history: the texts and the larger time-text, the communal narrative of which individual histories are expressions. This metahistorical time-text, far from being bound or limited by an objective, independent past-in-itself, is capable, as I will show, of extreme variation, of violent and irreconcilable change.

The study that follows is an attempt to describe two systems of historical thought considered as literary constructs, to account for the relatively sudden and violent transition between them, and to draw out some of the epistemological consequences of that transition for modernity. I have not attempted a comprehensive survey of historiography, for the result of comprehensiveness is an inevitable superficiality with regard to any particular author or history. Rather, I have selected certain nodes in the ideational development of the two historical models and examine these in detail. A certain arbitrariness of form results. Like lightning, which varies the direction of its travel with each new pulse, seeking the direction of least resistance, this study might have achieved its result through many alternative pathways. Such important figures in the development of the modern historical mind as Comte, Hegel, and Marx are bypassed, and considerable attention given to relatively obscure histories of Puritan New England because they serve as simplified and concentrated exemplars of the intellectual movements I am examining. One reason for the peculiarities of my choice is that the Enlightenment tradition of historiography has been admirably examined by Bury and by Koselleck.[6] Both trace the essential rupture between modern and premodern historical thought to Enlightenment natural philosophy: the Cartesian and the Hobbesian. I wish to push back the rupture, to give priority to sacred history over secular, to suggest that the movements described by Bury and Koselleck are reactions to and attempted compensations for a crisis that had already occurred.

I hope that this study is suggestive far beyond the texts it explicates; that those histories excluded from explicit mention will be in some way explained, even comprehended, by its paradigms; that, through this configuration of demonstrations as well as through any other, it will strike ground at last.

Los Angeles   A. K.
June 1990

Acknowledgments

Heartfelt thanks are due to all who read this work while it was being written and offered advice and encouragement: Sacvan Bercovitch, Amy Clampitt, Andrew Delbanco, Claudine Frank, Carl Hovde, and Blanford Parker; to Rebecca Monroe and Steve Novak, who treated me to dinner at the completion of every fifty pages; and to Harvard University, which afforded me employment, access to its libraries, and some very good friends.

Note on Texts

As none of my arguments depend on textual variation, convenience and availability, both for myself and for any reader who wishes to find the works quoted and mentioned here, have been almost the sole criteria for my choice of texts. All foreign-language works are given both in the original and in English translation. I have tried to point the reader to complete translations and to modern editions, where these exist; where they do not, I have resorted to my own translations and to early editions.

Contents

The Estrangement of the Past

Not to begin to be; and so not be limited by the concernments of *Time,* and *Place;* is the Prerogative of GOD alone. But as it is the Priviledge of Creatures, that GOD has given them a beginning: so to deny their *actions,* or *them,* the respect they bear to *Place,* and *successive duration,* is, under a pretence of Promotion, to take away their very Being.

SAMUEL SEWALL, *Phaenomena quaedam Apocalyptica*
ad Aspectum Novi Orbis configurata. Or, Some Few
Lines towards a Description of the New Heaven as It
Makes to Those who Stand upon the New Earth
(Boston, 1697)

1

Foundations of the Medieval Past

> Quelle plus grande abdication de Dieu que le temps? Nous
> sommes abandonnés dans le temps. Dieu n'est pas dans le temps.
>
> (What greater abdication of God than time? We are abandoned
> in time. God is not in time.)
>
> —SIMONE WEIL, *La Connaissance surnaturelle*

If the Gospels and the Acts are excluded, the first attempt at a Christian history was the *Chronographia* of Sextus Julius Africanus (written 212–221), of which only fragments are extant.[1] The *Chronographia* is not a history in the sense of a detailed and interpretive narrative, but a chronological table. It seeks to reconcile what, before it, had been two generally antagonistic accounts of the past: the salvific history of the Jews, a progression of manifestations of God's grace, which in retrospective Christian interpretation lead to the messianic Advent of Christ; and the imperial Greco-Roman history, in which all the cultures of Greece and Asia prophesy of and flow toward their highest manifestation in imperial Rome, the fulfillment both of Athens' and of Troy's promise. Of this imperial historical ideology the *Aeneid* is perhaps the most complete expression. Africanus's *Chronographia* brings these two mutually uncomprehending histories together in outline. The great events from which dates are calculated are the Creation, the Flood, *and* the Olympiads. Although concerned primarily with the syncretic dating of all received history, and not with narrative or interpretation as such, its very act of syncretism was interpretive.

3

The *Chronographia* exerted enormous influence on all subsequent Western histories: it was revised and expanded by Eusebius of Caesarea (his *Chronicorum libri duo*), which was in turn revised and expanded by Jerome, then by Prosper of Aquitaine, Isidore of Seville, and others.[2] With the *Chronographia*, Christian historiography begins in an act of harmonization, of syncretization. Given two incompatible versions of the past, what is irreconcilable must be reconciled, made up into a single fabric. The reconciliation in the *Chronographia* is only in the matter of dates, but this precedent is important nonetheless. The two traditions could now be comprehended in simultaneity, seen together on the same page.

Eusebius was the first continuator of Africanus, and it was really Eusebius's *Chronicorum* that became the model for all medieval chronology, and indeed of all universal chronologies down to the present. In addition to a seamless chronology of the universal past, Eusebius also attempted a more difficult task: a seamless and complete history of the Church of Christ down to his own day, and this not merely a description of events and dates, but an interpretation of what the Church's three hundred years of persecution and waiting meant. Eusebius's *Historia ecclesiastica* is the first narrative, interpretive Christian history (both his and Africanus's chronicles consisted only of tables), and, together with the voluminous historical writings of Augustine, it builds the wall around the city. Eusebius and Augustine dictate the form and the limits of medieval historical thought; they are the architects and topographers of that city's reach; everything that follows them is domestic architecture and the life lived within it.

Eusebius's *Historia* attempts a comprehensive telling of the history of the Church, from Christ's Advent to Eusebius's present, and as such it must contain, at least implicitly, an interpretation of the *structure* of time. Eusebius could have been a diarist or a witness, rather than a historian; that is, he could have related a few specific events, given a few martyrologies, and become immersed in the detail of the story he was telling. This is what most of his sources had done; but, instead, he set himself the task of compiling all of the known sources. It is precisely this matter of comprehensiveness or comprehension (in all of this word's senses) that differentiates the historian from the source. The longer the period dealt with, the greater the accumulation of details, the stronger becomes the need to see movements, patterns, tendencies, meanings in the story being told. Therefore, as the first comprehensive chronicler of the Church, Eusebius *had* to formulate comprehensively a historical structure that had been developing in parts and fragments throughout the New Testament books and in the writings of the pre-Eusebian church fathers.

Christianity suffers from an essential historical problem, in that it is a religion based on historical events. The Incarnation of God takes place in the particular man Jesus, at a particular time; it is not repeated in each new generation. All religions posit some kind of separation between the believer and God; indeed, religions and myths are grammars of absence; they are complex verbal formulas that bring into consciousness that which is beyond the limits of phenomenal perception: the origins and *telos* of the universe, survival beyond death, the transcendent. In most religions this separation is phenomenal or perceptual; in Christianity it is also temporal. For the Church of the second and third centuries, there must have arisen a very real psychological crisis as to the perceptual object of faith. Christ the God-in-man has ascended, according to the apostolic witnesses. The object of faith, then, cannot be Christ himself, but the memory of Christ preserved by these witnesses; but the witnesses, although delivered from death in a spiritual or metahistorical sense, all died within history. The result was the inevitable reification of the living person of Christ into a text (actually a whole series of uncertain and competing texts, some outright forgeries), which became the object of faith when no living witnesses remained. With each passing generation, the person of Christ became more absent, more tenuous. To counteract the withdrawal of the historical Christ, believers had the words of Christ concerning the Spirit[3] and the Second Coming:[4] the present and the future promises. But both of these comforters are imbued with severe problems. The experience of the Spirit is necessarily mystical and enthusiastic in nature (in enthusiasm's original sense of being possessed by a god). Anyone can lay claim to revelations from the Holy Spirit, and there is no way to test such revelations other than by a written and standard body of doctrine, the words of Christ and of the apostles embodied in fixed texts. The Church was rife with a multiplicity of warring interpretations within its first generation, each neologism of faith or practice claiming the authority of the Spirit. John[5] writes:

> Ἀγαπητοί, μὴ παντὶ πνεύματι πιστεύετε, ἀλλὰ δοκιμάζετε τὰ πνεύματα εἰ ἐκ τοῦ θεοῦ ἐστιν, ὅτι πολλοὶ ψευδοπροφῆται ἐξεληλύθασιν εἰς τὸν κόσμον.

(Beloved, do not believe every spirit, but test the spirits to see whether they are of God; for many false prophets have gone out into the world.)[6]

And in order to test, one must have a fixed, orthodox, written creed. The more the Spirit is invoked, the more must the Church turn back to the text.

The Second Coming is equally problematic as a focus for belief. Its promise is dependent on Christ's words, which are in turn dependent on the certainty or uncertainty of the written text, their inseparable matrix; and textual uncertainty was very real to the early Church. Its authority to create a canon was the retrospective judgment of a later age: the creators themselves labored over the genuineness—the apostolic authorship or authorship under apostolic auspices—of the books they admitted. The uncertainty might be gauged from Eusebius's list of the New Testament books (written sometime between 300 and 330), in which he doubts the authorship of Revelation and of Second and Third John.[7] Even beyond the textual problems, the eschatological passages are difficult to interpret. John seems to have believed that Christ's return was imminent:

Παιδία, ἐσχάτη ὥρα ἐστίν, καὶ καθὼς ἠκούσατε ὅτι ἀντίχριστος ἔρχεται, καὶ νῦν ἀντίχριστοι πολλοὶ γεγόνασιν· ὅθεν γινώσκομεν ὅτι ἐσχάτη ὥρα ἐστίν.

(Children, it is the last hour; and as you have heard that antichrist is coming, so now many antichrists have come; therefore we know that it is the last hour.)[8]

John believed that Christ was coming soon, at any moment, not in millennia. As time accumulated, as the Church developed a history, the living Christ receded into the past, beyond all but written memory, and receded into the future as the awaited return did not come. This time of no revelation, of God's silence, of the long waiting, of the last hour that stretches into infinity—this was the history that Eusebius had to invent a shape for. In the terrible paradox of faith, of the blessing and curse of believing what has not been seen, the people whose city was not of this world had to prepare an earthly habitation for itself: an organized Church and an ideology of time; and both of these are substitutes for the immediate presence of the Savior one was born too late to know, too early to see come in glory. In Christian consciousness—and the consciousness of the West cannot be separated from a specifically Christian consciousness—history is the absence of God. Everything that follows is both denial and confirmation of this.

In the first book of his history, Eusebius lays out a statement of purpose that is in itself a philosophy of time:

Τὰς τῶν ἱερῶν ἀποστόλων διαδοχὰς σὺν καὶ τοῖς ἀπὸ τοῦ σωτῆρος ἡμῶν καὶ εἰς ἡμᾶς διηνυσμένοις χρόνοις, ὅσα τε καὶ πηλίκα πραγματευθῆναι κατὰ τὴν ἐκκλησιαστικὴν ἱστορίαν λέγεται. . . .

(I have purposed to record in writing the successions of the sacred apostles, covering the period stretching from our Saviour to ourselves; the number and character of the transactions recorded in the history of the Church. . . .)[9]

These "successions" (διαδοχαί) suggest a passing of the mantle from one generation to the next, that the authority of those first witnesses has not diminished. The Greek word contains no suggestion of permutation or development, but rather of sameness. As the translator whose English text I have used above puts it in his commentary on the word:

"Succession," . . . did not merely mean, though it certainly included, the apostolic succession of the bishops of the four great "thrones," but rather the whole intellectual, spiritual, and institutional life of the Church. It cannot be too strongly emphasized that Eusebius, like all early church historians, can be understood only if it be recognized that whereas modern writers try to trace the development, growth, and change of doctrines and institutions, their predecessors were trying to prove that nothing of the kind ever happened. According to them the Church had one and only one teaching from the beginning; it had been preserved by the "Succession" and heresy was the attempt of the Devil to change it.[10]

A word in this first passage of Eusebius deserves special mention: πραγματευθῆναι, translated here as "transactions." A very general term meaning affairs, pursuits, actions, deeds, matters, things, tasks, disputes, or lawsuits, it is very definitely not a spiritual term. Its verb form, πραγματεύομαι, means to trade or do business. The point is that these are the very earthly acts, pragmata, of a Church exiled in "the period stretching from our Saviour to ourselves." In this time of silence between divine events, faith and authority can only be founded on the past: Christ and the apostles, who recede each year further from the grasp of present knowledge, into the ungraspable tenuousness of collective memory. The only way to palliate this recession of the focus of faith is to deny psychologically, and subsequently ideologically, the mutability of time and, as mutability is the essence of our experience of time, consequently to deny time itself. Eusebius continues with his list of subjects:

τίνες τε καὶ ὅσοι καὶ ὁπηνίκα νεωτεροποιίας ἱμέρῳ πλάνης εἰς ἔσχατον ἐλάσαντες, ψευδωνύμου γνώσεως εἰσηγητὰς ἑαυτοὺς ἀνακεκηρύχασιν, ἀφειδῶς οἷα λύκοι βαρεῖς τὴν Χριστοῦ ποίμνην ἐπεντρίβοντες. . . .

(the names, the number and the age of those who, driven by the desire of innovation to an extremity of error, have heralded themselves as the intro-ducers of Knowledge, falsely so-called, ravaging the flock of Christ unspar-ingly, like grim wolves.)[11]

Heresy comes from innovation. Lacking the immanent presence of God-among-us, Eusebius has instead the text, the only manifestation of God available to his generation. Given the increasingly extending withdrawal of Christ from history, into past and into future, the Church must remain faithful to the text and its orthodox interpreta-tion, which within Eusebius's rhetoric is made original, not itself the subject of development. This is, of course, an ingenious fiction, although far repressed from consciousness as such. It ignores the Church's creation of those texts (the canon was still very much in dis-pute at the time Eusebius wrote) and the development of doctrine up to the first Council of Nicea in 325, at which orthodoxy was hacked out of chaos, and bitter cliques feuded over distinctions of doctrine that cannot be separated from the exquisite semantic quiddities of the Greek in which they were couched. Later councils reversed earlier decisions, and even Eusebius was not always on what the Church would take to be the orthodox side. Some manuscripts of the *Historia* even contain an interpolation appended to the table of contents of the first book, warning the reader that the author is a heretic.[12]

Against this chaotic and dynamic development of orthodoxy, Euse-bius opposes an ontological fiction. His history proposes a literary, conceptual form imposed on time: Christ has ascended; his presence remains in his *words*, as reported faithfully in the Gospels and inter-preted correctly in the writings of the apostles; the locus of belief is therefore in the past; no new truth will be given or discovered until Christ returns; therefore, any innovation will be a denial of Christ's teaching. The effect of this verbal or literary form imposed on the chaos of actual events—and the difficulty of even knowing what those events are, once they have receded from "living" memory—is to annul, by means of words, the psychological experience of time and change so that the present, again by means of a text or rhetoric, can be united psychologically with the past. The rhetoric of temporal union becomes an anodyne for the experience of those canceling deaths and generations between present and past; it is the necessary story that denies the abyss.

Even Eusebius's myth of unchanging time was a danger to the unity it asserted, because it was itself an innovation. Here, at some length, is his apology for his own work; it conveys not only his anxiety at his own innovation, but also the aching sense of loss that underlies and necessitates the rhetoric of union:

ἀλλά μοι συγγνώμην εὐγνωμόνων ἐντεῦθεν ὁ λόγος αἰτεῖ, μείζονα ἢ καθ᾽ ἡμετέραν δύναμιν ὁμολογῶν εἶναι τὴν ἐπαγγελίαν ἐντελῆ καὶ ἀπαράλειπτον ὑποσχεῖν, ἐπεὶ καὶ πρῶτοι νῦν τῆς ὑποθέσεως ἐπιβάντες οἷά τινα ἐρήμην καὶ ἀτριβῆ ἰέναι ὁδὸν ἐγχειροῦμεν, θεὸν μὲν ὁδηγὸν καὶ τὴν τοῦ κυρίου συνεργὸν σχήσειν εὐχόμενοι δύναμιν, ἀνθρώπων γε μὴν οὐδαμῶς εὑρεῖν οἷοί τε ὄντες ἴχνη γυμνὰ τὴν αὐτὴν ἡμῖν προωδευκότων, μὴ ὅτι σμικρὰς αὐτὸ μόνον προφάσεις, δι᾽ ὧν ἄλλος ἄλλως ὧν διηνύκασι χρόνων μερικὰς ἡμῖν καταλελοίπασι διηγήσεις, πόρρωθεν ὥσπερ εἰ πυρσοὺς τὰς ἑαυτῶν προανατείνοντες φωνὰς καὶ ἄνωθέν ποθεν ὡς ἐξ ἀπόπτου καὶ ἀπὸ σκοπῆς βοῶντες καὶ διακελευόμενοι, ᾗ χρὴ βαδίζειν καὶ τὴν τοῦ λόγου πορείαν ἀπλανῶς καὶ ἀκινδύνως εὐθύνειν. ὅσα τοίνυν εἰς τὴν προκειμένην ὑπόθεσιν λυσιτελεῖν ἡγούμεθα τῶν αὐτοῖς ἐκείνοις σποράδην μνημονευθέντων, ἀναλεξάμενοι καὶ ὡς ἂν ἐκ λογικῶν λειμώνων τὰς ἐπιτηδείους αὐτῶν τῶν πάλαι συγγραφέων ἀπανθισάμενοι φωνάς, δι᾽ ὑφηγήσεως ἱστορικῆς πειρασόμεθα σωματοποιῆσαι, ἀγαπῶντες, εἰ καὶ μὴ ἁπάντων, τῶν δ᾽ οὖν μάλιστα διαφανεστάτων τοῦ σωτῆρος ἡμῶν ἀποστόλων τὰς διαδοχὰς κατὰ τὰς διαπρεπούσας ἔτι καὶ νῦν μνημονευομένας ἐκκλησίας ἀνασωσαίμεθα. ἀναγκαιότατα δέ μοι πονεῖσθαι τὴν ὑπόθεσιν ἡγοῦμαι, ὅτι μηδένα πω εἰς δεῦρο τῶν ἐκκλησιαστικῶν συγγραφέων διέγνων περὶ τοῦτο τῆς γραφῆς σπουδὴν πεποιημένον τὸ μέρος. . . .

(Even at that point the project at once demands the lenience of the kindly, for confessedly it is beyond our power to fulfil the promise, complete and perfect, since we are the first to enter on the undertaking, as travellers on some desolate and untrodden way. We pray God to give us his guidance, and that we may have the help of the power of the Lord, for nowhere can we find even the bare footsteps of men who have preceded us in the same path, unless it be those slight indications by which in divers ways they have left to us partial accounts of the times through which they have passed, raising their voices as a man holds up a torch from afar, calling to us from on high as from a distant watch-tower, and telling us how we must walk, and how to guide the course of our work without error or danger. We have therefore collected from their scattered memoirs all that we think will be useful for the present subject, and have brought together the utterances of the ancient writers themselves that are appropriate to it, culling, as it were, the flowers of intellectual fields. We shall endeavour to give them unity by historical treatment, rejoicing to rescue the successions, if not of all, at least of the most distinguished of the apostles of our Saviour throughout those churches of which the fame is still remembered. To work at this subject I consider especially necessary, because I am not aware that any Christian writer has until now paid any attention to this kind of writing. . . .)[13]

The "desolate and untrodden way" is both the path of innovation and the journey of history from Christ's time until Eusebius's. To chart and make domestic this wilderness, he has only "slight indications," "partial accounts," "scattered memoirs." When examined, the textual sources of any historical narrative are shockingly fragmentary, and behind them lies some unrecoverable reality that can only be approached, but never possessed. What haunts Eusebius in this passage is the idea that the past, even a past more recent than the time of Christ, may be unknowable, not only in its larger meanings, but also in its details. If history is unknowable, then so is Christ. The history Eusebius writes is not given, though it proposes a doctrine of givenness; it must be wrested from a dark, waste past in an *agonia* of active verbs: "to give them unity by historical treatment . . . to rescue the successions"; and these not of all, but only of the most distinguished, of the apostles and churches "of which the fame is still remembered."

Rhetoric, of which history is one subgenre, is the linguistic assertion of consciousness. Words are signifiers of absent things—otherwise the sign would not be needed—and it is precisely the experience of absence that tends to assert presence in words, in myth and rhetoric and doctrine. Words substitute for the impalpable. This is not to say that the impalpable is not real, or that a knowledge of it arrived at by linguistic discourse is not a real knowledge; indeed such a linguistic perception may be more real than any empirical perception, but it cannot be sure. This is the perception of faith, and faith, if it is intellectually honest, is a conclusion including doubt, and always necessitated by the absence of sensory experience. So it is the absence, or at least uncertainty, of affectional bonds that leads to a formulated sentimentality; the experiential absence of God that motivates a verbal theology; the absence of the past that demands a written history. Language, and its ability to evoke, fills an experiential void, and this evocation is both ambiguous and dangerous. Perhaps the evocative abilities of language do bring about a genuine presence: the mystical experiences of a Saint Theresa or a Saint John of the Cross cannot be separated from their verbal formulation. Another possibility, though, is that, just as lies cannot be distinguished from language, so perhaps language cannot be distinguished from lies. The very act of laboriously formulating consciousness into self-description may be autochthonous to an original lack of consciousness, language perhaps an illusory means of bringing into being things that are not. Neither interpretation necessarily excludes the other; they can coexist in infinitely complex permutations, according to one's faith. Eusebius's formulation of history is such a complex interplay of absence and presence. The rhetoric is all of presence, of a unity with the past and with Christ; beneath can be discerned the terrible fear of discontinuity, the void his words must fill. His need to make time uniform, as the medium through which human faith must travel to its object, is so extreme that not only is his own innovation a threat, but also the innovation of the Incarnation:

> ταύτῃ δὲ καὶ τῆς Χριστιανῶν ἀρχαιότητος τὸ παλαιὸν ὁμοῦ καὶ θεοπρεπὲς τοῖς νέαν αὐτὴν καὶ ἐκτετοπισμένην, χθὲς καὶ οὐ πρότερον φανεῖσαν, ὑπολαμβάνουσιν ἀναδειχθήσεται.

> (By this means, moreover, the real antiquity and divine character of Christianity will be equally demonstrated to those who suppose that it is recent and foreign, appearing no earlier than yesterday.)[14]

He accomplishes this by reading every manifestation of God in the Old Testament as being very Christ:

> καὶ μὴν οὐδ᾽ ὑποβεβηκότων ἀγγέλων καὶ λειτουργῶν θεοῦ τὰς ἀναγραφείσας θεοφανείας ὑπονοεῖν θέμις.

(And it cannot be right to suppose that the Theophanies described were the appearances of subordinate angels and ministers of God.)[15]

Eusebius reads Christ back into more than the messianic prophecies identified as such by the Gospel writers. Each plural pronoun applied to God, each angelic visitation, each use of "Lord" is taken to mean Christ instead of the Father. Even the word "word" in Hebrew is interpreted by Eusebius, with the convenient interposition of the Septuagint, to denote the specifically Greek concept of *Logos*. The result is to make the Incarnation less important: Christ becomes literary, transcendent, ahistorical, the eternal *Logos* rather than the man Eusebius and his generation were born too late to see.

The elimination of historical or mutable time, then, acts as a comforter for those deprived of the direct revelation. Eusebius does of course give a history of events and generations, but his theoretical rhetoric works against the accumulation of years that he chronicles. If the mutable and innovative character of history can be denied, then the believer can achieve a rhetorical, linguistic union with the past, and a linguistic union can become a mystical union. The denial of mutability has a second, related, and essential function: as the authority of truth lies in the past only, any mutation from that past will transform the content of Christian belief, by slow degrees, into something unrecognizable. This was especially true in a period in which the Church was split into factions of orthodoxy and heterodoxy. What Eusebius cannot afford to admit is that his own positions in the various doctrinal disputes are as much refinements and developments of Christ's and of the apostolic teaching (which, to add complexity, was itself a refinement and development of Christ's teaching, whatever that was apart from the matrix of the apostolic texts) as any of the positions of the heterodox. Each position claiming to be orthodox must also claim absolute uniformity with the past, else the whole temporal fabric of faith will unravel, and all positions will become equally neologisms. What Eusebius fears is an Ariadne's thread far too complex to be retraced—a labyrinth of time with convolutions and corners that cut off forever the prospect of where one has been, that leads ever and only inward, away from the light of the divine event and into an interiority of fragmented history. The void that lies behind his rhetoric of union is not so much the impossibility of faith in a history so conceived, as its complexity. Thus, the past must be simplified: the Church was virgin until the death of the apostles:

> ... ἐπιλέγει ὡς ἄρα μέχρι τῶν τότε χρόνων παρθένος καθαρὰ καὶ ἀδιάφθορος ἔμεινεν ἡ ἐκκλησία, ἐν ἀδήλῳ που σκότει ὡς εἰ φωλευόντων εἰς ἔτι τότε τῶν, εἰ καί τινες ὑπῆρχον, παραφθείρειν ἐπιχειρούντων τὸν ὑγιῆ κανόνα τοῦ σωτηρίου κηρύγματος· ὡς δ᾽ ὁ ἱερὸς τῶν ἀποστόλων χορὸς διάφορον εἰλήφει τοῦ βίου τέλος παρεληλύθει τε ἡ γενεὰ ἐκείνη τῶν αὐταῖς ἀκοαῖς τῆς ἐνθέου σοφίας ἐπακοῦσαι κατηξιωμένων, τηνικαῦτα τῆς ἀθέου πλάνης ἀρχὴν ἐλάμβανεν ἡ σύστασις διὰ τῆς τῶν ἑτεροδιδασκάλων ἀπάτης, οἳ καὶ ἅτε μηδενὸς ἔτι τῶν ἀποστόλων λειπομένου, γυμνῇ λοιπὸν ἤδη κεφαλῇ τῷ τῆς ἀληθείας κηρύγματι τὴν ψευδώνυμον γνῶσιν ἀντικηρύττειν ἐπεχείρουν.

(... until then the church remained a pure and uncorrupted virgin, for those who attempted to corrupt the healthful rule of the Saviour's preaching, if they existed at all, lurked in obscure darkness. But when the sacred band of the Apostles and the generation of those to whom it had been vouchsafed to hear with their own ears the divine wisdom had reached the several ends of their lives, then the federation of godless error took its beginning through the deceit of false teachers who, seeing that none of the Apostles still remained, barefacedly tried against the preaching of the truth the counter-proclamation of "knowledge falsely so-called.")[16]

The solution is to hold to the past, to the written word:

> ... ἐν πρώτοις μάλιστα προφυλάττεσθαι τὰς αἱρέσεις ἄρτι τότε πρῶτον ἐπιπολαζούσας παρῄνει προύτρεπέν τε ἀπρὶξ ἔχεσθαι τῆς τῶν ἀποστόλων παραδόσεως, ἣν ὑπὲρ ἀσφαλείας καὶ ἐγγράφως ἤδη μαρτυρόμενος διατυποῦσθαι ἀναγκαῖον ἡγεῖτο.

(He [reference uncertain, either Papias or Ignatius] particularly warned them to be on their guard against the heresies which then for the first time were beginning to obtain, and exhorted them to hold fast to the tradition of the Apostles, to which he thought necessary, for safety's sake, to give the form of written testimony.)[17]

The search for union with the past led to a cult of elders (those who had actually met an apostle);[18] when this was no longer possible, to a fetishization of the written record; and later to the reverence of relics. The function of all of these is identical to that of the icon: they provide tangible objects of contemplation that may lead beyond themselves to the intangible. But the iconographic imagination is fraught with two perils: the icon may itself become the object, in which case the intangible beyond disappears and the representation subsumes the represented into itself; or, the icon comes to represent the absence of the represented, the absence that necessitates the icon. Christian icons represent a double absence: they point to the intangible beyond the world, the risen Christ and the eternal God; and also the intangible past, Christ on earth. Each icon, then, is both the invocation of presence, and the declaration of the absence that requires an invocation. Faith generally declares the presence and sublimates the absence, which is the hidden meaning of the icon that separates, as the iconostasis, the screen hung with icons, of the Eastern Orthodox Church separates the congregation from the *bema* (sanctuary), the place of the (invoked) presence. Eusebius's historical form is just such an iconostasis; through its doctrine of union—the denial of difference or discontinuity within time—it provides a matrix for all the icons, a setting in which the jewels so dear to medieval contemplation could be embedded. Without such a setting, such an overarching scheme, the martyr's relics, the apostles' words, the real presence in the sacraments, the inherited authority of the Church would not cohere; the mosaic built on Eusebius's past would be patternless. Yet while the icons unite, they also separate. The very strength of the declaration, of its intellectual effort, reveals the possibility of incoherence. This is the darkness couched beneath.

In his introduction to *L'Archéologie du savoir*, Michel Foucault characterizes the new schools of history by their willingness to perceive difference and discontinuity in time; previously all Western history has sought continuity and totality:

> L'histoire continue, c'est le corrélat indispensable à la fonction fondatrice du sujet: la garantie que tout ce qui lui a échappé pourra lui être rendu; la certitude que le temps ne dispersera rien sans le restituer dans une unité recomposée; la promesse que toutes ces choses maintenues au loin par la différence, le sujet pourra un jour—sous la forme de la conscience historique—se les approprier derechef, y restaurer sa maîtrise et y trouver ce qu'on peut bien appeler sa demeure.

> (Continuous history is the indispensable correlative to the founding function of the subject: the guarantee that everything that has eluded may be

restored to him; the certainty that time will disperse nothing without restoring it in a reconstituted unity; the promise that one day the subject—in the form of historical consciousness—will once again be able to appropriate, to bring back under his sway, all those things that are kept at a distance by difference, and find in them what might be called his abode.)[19]

Foucault—who also designates this continuous history as "total history" as opposed to "general history"—is aware of the consolatory function of this mode of historical thought. I am approaching the same distinction from the other end of its time line. Eusebius is the first elucidator of continuous and total history in its peculiarly Western sense. Classical history had its own forms of totality and consolation, but never before had there been such a peculiar imperative to overcome the separations of time. To keep the world in touch with its monochronically incarnated savior required a radical denial of the movement, the discontinuity of time. This is the structural distinction of pre-Reformation Christian historiography, and Eusebius— although he transmits the developing and fragmentarily expressed doctrine of a whole community—is its first clear formulator. In terms of extant texts, the *Historia ecclesiastica* is the point at which Foucault's "total history"—from which he sees postmodernism attempting its escape—enters the Western mind.

There are two more observations to be made before leaving Eusebius's *Historia ecclesiastica*, almost in passing at this point, but which will take on great importance in the light of what is to come.

First, the bulk of the *Historia* is taken up with martyrology, hideously explicit and piteously affecting. The early Christians endured a holocaust (exceeding those of our own century in its aesthetic variations of cruelty, although not in systematic efficiency of slaughter) that lasted not three years, but almost three hundred. This was a people in need of consolation. Martyrdom is treated as a sacrament by Eusebius and his parade of witnesses, more precious and efficacious even than communion and baptism. Martyrdom, the keeping of faith at the extremity of torture, completes faith, is a gift of power from God. It is a point of mystical extremity, a breaking through the fabric of the world into the presence of God. Like the martyrdom of Christ, each imitative martyrdom has that double tincture of substance and absence in that the victim's abandonment by God is also the moment of greatest union with God, by extremity of faithfulness when there is least cause for faith, and by mystical participation in the death of Christ. The odor of blood becomes the odor of sanctity.

Second, Eusebius continues the syncretism of Africanus's and his own chronological tables by uniting the imperial history with the salvific history. The Roman government is treated throughout as a just government, ordained by God. The persecutors of the Church, including the persecuting emperors, are presented as aberrations. To give one example, Eusebius relays the fantastic tradition that Pilate sent a report, credulous of Christ's divinity, to the senate, and that Christ narrowly escaped election by this body to the Roman pantheon, on a technicality. Both Pilate and the emperor Tiberius are depicted as believers.[20] This is indicative not only of how much the early Church tried to ingratiate itself with the Roman state in all things permissible, but also of an aching sense of cultural identity with and loyalty to the institution that persecuted it. The inherently institutional nature of that violence is denied in Eusebius. Evil is attributed only to individuals, including the persecuting emperors, within an otherwise just institution; to the subject peoples of the empire, who instigate more persecutions than the Romans themselves; and particularly to the Jews, who are the consistent enemies of Christ and deserve all the punishments the Romans, as the agents of divine retribution, inflict upon them.[21] In the interstice between the New Testament books and Eusebius, Christianity had changed from cultural Judaism, accepting gentiles only with difficulty, to a thoroughly gentile, Romanized, and anti-Jewish movement.

These two elements, martyrology and cultural devotion to Rome, dictate the "plot" or narrative structure of the *Historia*. Unfortunately, it does not lend itself to demonstration by quotation; the movement is simply too large, and can only be discerned by reading the *Historia* in its entirety. The narrative begins with the Incarnation and Atonement, and the evangelical witness of the apostles. These are acts of integration of God and mankind, and provide the *Historia* with that initial unity from which narratives and histories commence. Then comes, after the first generation, disintegration as heresies multiply and the Church falls under persecution. This is the time of separations, when faith is validated precisely as faith, when the glory of the Church consists in its assertion of the faithfulness of God among the darkness and blood of its apparent abandonment by God. The *Historia* concludes with a reintegration, a very real deliverance brought about not by the Second Coming of Christ, but by the emperor Constantine. The accession of Constantine, "the friend of God," is the *telos* of the plot, resolving all conflicts. The tenth book of the *Historia* (again, not really conveyable by quotation) is taken up by an extraordinary messianization of the emperor. In Eusebius's rhetorical imagination and in the structure of the *Historia*, the coming of Constantine and his righting of all wrongs for the Church, perhaps the most unforeseeable and unlooked-for event in all of Western history, has taken the place of the Second Coming of Christ.

The consistent tendency of apocalyptic groups, when the promised divine event fails to occur, is to reconceptualize prophecy so that the event can be said to have occurred, either spiritually or secularly: Christ has returned, but in spirit, and can only be perceived in spirit, by the spiritual.[22] By spiritualization, which is conversely an act of secularization, the miracle becomes untestable, demiracularized. The analogue between the early Church and later apocalyptic groups is a very limited one: these groups have specified dates for the Second Coming, which has required some very rapid and agile prevarication on the day after; the early Church had instead an indefinite promise, but one whose preconditions seemed to have been fulfilled by the time of Titus's sack of Jerusalem in A.D. 70. A spiritualizing/secularizing tendency seems to have arisen even before this: something of the kind must have happened in the church of Thessalonike, a general belief that Christ had already come, but invisibly,[23] not as the lightning shines from the East to the West.[24] Paul's answer to this gnostic return—perceptible only by the enlightened, one assumes—is that the antichrist has not yet appeared; therefore, the hour has not yet come. This contradicts John's opinion that the prophecies of antichrist had been fulfilled and that the Second Coming was immanent,[25] and indicates the interpretive difficulties even among the apostles themselves. In Eusebius there is not so much a spiritualization as a rhetorical substitution. Of course Constantine is not Christ, but he is the divine deliverer, and he closes the plot of persecution, abandonment, and absence in Eusebius's Christian history. The tenth book of the *Historia* proposes a closure to the openness of history and suffering; it ends the ache of longing for the return that, generation after generation, does not come, and the tense paradox of living in the time-bound human mire but believing oneself, against all the evidence of the senses, an immortal citizen of heaven. In a sense, after Constantine,

or at least after Eusebius's Constantine, there is no need for Christ to
return: the Church has an earthly city.

As Eusebius became almost the sole arbiter of ecclesiastical history for
the Middle Ages, so Augustine became the arbiter of universal history.
If Eusebius's *Historia* is a "total" history in Foucault's distinction,
then Augustine's, in his various historical works, is more so. Eusebius
formulated, codified the Christian interpretation of history from
Christ to Constantine; Augustine from the Creation to the end of the
world. His influence, like that of Eusebius, was, for the pre-Reformation West, mythopoeic; they invented the world of conceptions that
comprehended their followers and through which these followers comprehended the world. The influence of Eusebius and Augustine on that
intellectual world might be compared to the influence of Marx and
Freud on our own: we have no language in which to debate them that
they have not given us.

   Apart from the Bible, Augustine used two principal sources: Varro's
De gente populi Romani and Jerome's *Chronicle*, which is a Latin
translation and continuation of the *Chronicorum libri duo* of Eusebius
(this is now extant in its entirety only in Armenian translation, so
thoroughly did each stage of the chronicle tradition subsume its antecedent), in turn derived from the *Chronographia* of Sextus Julius Africanus (extant only in fragments). To this syncretic outline of Romano-
pagan and Jewish history, Augustine gave vast and systematic form. If
the line of influence is traced in the other direction, it leads to Isidore's
two histories, Bede's *De temporum ratione*, and Hugo of St. Victor's
De tribus maximis circumstantiis gestorum.[26] All of these are direct
imitations of Augustine's sixfold exposition of universal history.

   For Augustine, the critical determinant of a Christian ideology of
time is the uniqueness and unrepeatability of Christ's sacrifice. All of
time is changed by that atonement, and in none of its aspects can the
world revert to the condition it suffered before. Having conquered
death, Christ can never again be subject to death; nor can believers,
once regenerate, delivered from sin, be ever again unregenerate, subject to sin:

> Absit autem a recta fide ut his Salomonis verbis illos circuitus significatos
> esse credamus quibus illi putant sic eadem temporum temporaliumque
> rerum volumina repeti ut, verbi gratia, sicut isto saeculo Plato philosophus

in urbe Atheniensi et in ea schola quae Academia dicta est discipulos docuit, ita per innumerabilia retro saecula multum quidem prolixis intervallis, sed tamen certis, et idem Plato et eadem civitas et eadem schola idemque discipuli repetiti et per innumerabilia deinde saecula repetendi sint. Absit, inquam, ut nos ista credamus. Semel enim Christus mortuus est pro peccatis nostris; *surgens* autem *a mortuis iam non moritur, et mors ei ultra non dominabitur;* et nos post resurrectionem semper cum Domino erimus.

(But heaven forbid that our true faith should allow us to believe that these words of Solomon denoted those cycles in which, as those others think, the same measures of time and the same events in time are repeated in circular fashion: on the basis of this cyclic theory, it is argued, for example, that, just as in a certain age the philosopher Plato taught his students in the city of Athens and in the school called the Academy, so during countless past ages, at very prolonged yet definite intervals, the same Plato, the same city, and the same school with the same students had existed again and again, and during countless ages to come will exist again and again. Heaven forbid, I repeat, that we should believe that. For Christ died once for our sins, but "rising from the dead he dies no more, and death shall no longer have dominion over him"; and after the resurrection "we shall always be with the Lord.")[27]

This excludes the most pervasive classical historical theory, the cyclical, and Augustine sets about its defeat with characteristically voluminous thoroughness.[28] Augustine sweeps away the logic of classical history, both within the text called Augustine and within history itself—that is, within all the texts that follow him. Nietzsche has not succeeded in reviving it. I will not try to replicate or provide a précis of the arguments here; those from pagan presuppositions are persuasive in overcoming pagan objections, and those from Christian presuppositions are unanswerable.

From the exclusion of cyclicality from history, two logical consequences follow: that the world and the human race are not eternal, that human history has a limit; and that at least one radical, irreversible change, the Incarnation, has taken place within history. From the first of these comes the inevitable question: how old is the race of men? Augustine's answer is that the total of the ages in the genealogies of Genesis, bridging the gap between Eden and the earliest datable history in Africanus/Eusebius, yields a figure of less than six thousand years between Adam and Augustine's present. He is aware of the discrepancies between the figures in the Hebrew text, reproduced in Jerome's Vulgate, and those in the Septuagint, reproduced in the Old Latin Versions, but he finds these insignificant.[29] From the second of the two consequential propositions—that at least one radical change has taken place in history—also comes an inevitable question: have any other such changes occurred? From the Old Testament account, Augustine sees a number of such changes. These do not denote development in the modern historical sense, but abrupt changes in God's dealings with mankind. Each dispensation—the time between these alterations—has lasted about a thousand years. So, from Adam to Noah was the first millennium of human history, from Noah to Abraham the second, from Abraham to David the third, from David to the Babylonian Captivity the fourth, from the Babylonian Captivity to the preaching of John the Baptist the fifth, from John until the end, the present age of the Christian Church, the sixth. Thus, there are six millennia of human history corresponding to the six days of creation, and to the six ages of man.

I have presented the scheme in its logical sequence; Augustine nowhere does so, but everywhere alludes to it as a self-evident system. Its most concise presentation is found in the *Expositions on the Book of Psalms*,[30] the most exhaustive in *De civitate dei*.[31] This passage, from the latter, gives some idea of the system's typological density:

In quo articulus quidam factus est et exordium quodam modo iuventutis populi Dei; cuius generis quaedam velut adulescentia ducebatur ab ipso Abraham usque ad hunc David. Neque enim frustra Matthaeus evangelista sic generationes commemoravit ut hoc primum intervallum quattuordecim generationibus commendaret, ab Abraham scilicet usque ad David. Ab adulascentia quippe incipit homo posse generare; propterea generationum ex Abraham sumpsit exordium; qui etiam pater gentium constitutus est, quando mutatum nomen accepit. Ante hunc ergo velut pueritia fuit huius generis populi Dei a Noe usque ad ipsum Abraham; et ideo in lingua inventa est, id est Hebraea. A pueritia namque homo incipit loqui post infantiam, quae hinc appellata est quod fari non potest. Quam profecto aetatem primam demergit oblivio, sicut aetas prima generis humani est deleta diluvio. Quotus enim quisque est, qui suam recordetur infantiam?

Quam ob rem in isto procursu civitatis Dei, sicut superior unam eandemque primam, ita duas aetates secundam et tertiam liber iste contineat, in qua tertia propter vaccam trimam, capram trimam, arietem trimum et inpositum est legis iugum et apparuit abundantia peccatorum et regni terreni surrexit exordium, ubi non defuerunt spiritales quorum in turture et columba figuratem est sacramentum.

(With David a new era began, and one may say that the young manhood of the people of God commenced, since from the time of Abraham to this time of David they had passed through a period resembling adolescence. For it was not without reason that the evangelist Matthew listed the generations in such a way as to assign fourteen generations to this first period, namely that from Abraham down to David. It is true that a man can exercise his reproductive function from adolescence, and therefore the series of generations began with Abraham, who was also made the father of nations when his name was changed. Up to this time, then, there was a period resembling childhood for this race of God's people, from Noah to Abraham himself, and for this reason they are found to have had a language, namely Hebrew. For man begins to talk in his childhood, after the period of infancy, which is so called because it lacks the power of speech. And surely oblivion swallows up this first age, as the first age of mankind was swallowed up by the Flood. For how many men are there in a hundred who recall their own infancy?

Thus, in the progress of the city of God, as the previous book includes a single age, the first, so this book covers two ages, the second and the third. And in this third age, as indicated by the heifer of three years, the she-goat of three years and the ram of three years, not only was the yoke of the law imposed, but a multitude of sins emerged, and the first stages of the earthly kingdom arose, yet there was no lack of spiritual men who are symbolized by the turtle-dove and the pigeon.)[32]

The last millennium, the sixth day, is in its fourth century at the time of Augustine's writing, and nothing new will be revealed until the return of Christ, which will end earth and time, and initiate the seventh day, which has no evening, but is an eternal sabbath. The enigmatic millennium of the twentieth chapter of Revelation, Augustine identifies with the sixth millennium, the age of the Church, in which Satan is bound by its presence:

Mille autem anni duobus modis possunt, quantum mihi occurrit, intellegi: aut quia in ultimis annis mille ista res agitur, id est sexto annorum miliario tamquam sexto die, cuius nunc spatia posteriori volvuntur, secuturo deinde sabbato quod non habet vesperam, requie scilicet sanctorum quae non habet finem, ut huius miliarii tamquam diei novissimam partem quae remanebat usque ad terminum saeculi, mille annos appellaverit eo loquendi modo quo pars significatur a toto; aut certe mille annos pro annis omnibus huius saeculi posuit, ut perfecto numero notaretur ipsa temporis plenitudo. Millenarius quippe numerus denarii numeri quadratum solidum reddit. Decem quippe deciens ducta fiunt centum, quae iam figura quadrata, sed plana est; ut autem in altitudinem surgat et solida fiat, rursus centum deciens multiplicantur, et mille sunt.

(Now the thousand years may be understood in two ways, so far as I can see: either because this event takes place in the last thousand years, that is, in the sixth millennium, the latter parts of which are even now passing, as if it were a sixth day, to be followed by a Sabbath without an evening, which is the rest of the saints without an end, so that by that figure of speech which speaks of the whole, meaning a part, he calls the last part of the millennium, or day, which remained before the end of the world, a thousand years; or he at least used the thousand years as the equivalent of the whole period of this world's history, in order to indicate by a perfect number the fullness of time. For the number one thousand represents the cube of the number ten, inasmuch as ten times ten makes one hundred, a square, but a plane figure, while to make a solid figure the hundred is again multiplied by ten, and becomes a thousand.)[33]

This is total history indeed, and one question that must be asked about it—if an attempt (the achievement may not be possible) is to be made to understand the kind of mind that formulated, imitated, and received it—is, what is the relationship between the literary structure, namely the mental model embodied in the text, and the truth of history? For this is history as a self-conscious literary construct: Augustine has neither archaeological (again in Foucault's sense) nor historical method. His history is a hermeneutical exercise, a literary interpretation of literary texts. For a universal history, he has only three sources (apart from a few asides concerning Pliny and the pseudo-Sibylline oracles, these occupying only a few pages): the Bible, the Africanus/Eusebius/Jerome chronicle, and Varro. By comparison, Eusebius, with a much more limited task, has assembled a cloud of witnesses: early fathers, oral traditions, city archives. Both authors are aware, however, that they are dealing precisely with texts, not revealing a historical truth that can be uncovered through and behind the texts by a critical destructuring of them. Eusebius's task is to assemble his sources into a coherent whole; Augustine's is to perform a work of structural-literary analysis on his sources, to make explicit the struc-

tures implicit within them. The devotion to the text incipient in Eusebius is full-grown in Augustine. There is no longer the uncomfortable,
not quite closable interstice between the text and the world, the words
about Christ and Christ himself, which is evident, although veiled, in
Eusebius; in Augustine, the text has overcome the world, has sub-
sumed all idea of a nontextual reality. In part this is due to the accumulation of text on text. Whereas Eusebius had to create coherence
from fragments and an uncertain canon, Augustine can take Eusebius's effort as given history, and can regard both Old and New Tes-
taments as a single book, because for him they are physically a single
book in his own language: Jerome's Vulgate. Thus the assembling
efforts of each generation comfort the next, free it from the anxiety of
innovation. While the comfort afforded by the possession of a coherent text should not be underestimated, Augustine's textual faith is far
too complex to be accounted for by it. It is not in the least primitive
or credulous, but philosophical in origin and entirely self-conscious.
Its foundation and center can be found in the great discourse on the
nature of time, which concludes the *Confessions*.[34] It begins with a
consideration of the separation of past, present, and future in the
imagination, and the idea of duration. If only the present is empiri-
cally extant, then surely past and future can have no duration as they
are unexperienceable:

> quod enim longum fuit praeteritum tempus, cum iam esset praeteritum,
> longum fuit, an ante, cum adhuc praesens esset? tunc enim poterat esse
> longum, quando erat, quod esset longum: praeteritum vero iam non erat;
> unde nec longum esse poterat, quod omnino non erat.

> (For that past time that was long, was it long when it was already past, or
> when it was yet present? For then might it be long, when there was what
> could be long; but when past, it was no longer; wherefore that could not
> either be long, which was not at all.)[35]

Therefore, only the present can be said to have duration, but:

> an centum anni praesentes longum tempus est? vide prius, utrum possint
> praesentes esse centum anni. . . . si quid intellegitur temporis, quod in nul-
> las iam vel minutissimas momentorum partes dividi possit, id solum est,
> quod praesens dicatur; quod tamen ita raptim a futuro in praeteritum
> transvolat, ut nulla morula extendatur.

(Are an hundred years in present a long time? See first, whether an hundred years can be present. . . . If any instant of time be conceived which cannot be divided either into none, or at most into the smallest particles of moments; that is the only it, which may be called present; which little yet flies with such full speed from the future to the past, as that it is not lengthened out with the very least stay.)[36]

If then the present has no duration, whence comes the idea of duration? Augustine considers the possibility that time is merely the motion of the heavenly bodies, but decides that the opposite is true: movement can only be perceived by reference to some independent and preceding idea of time. Strangely perhaps, he finds his solution in poetry and the measurement of meter: mind and language are the arbiters of time:

et qui narrant praeterita, non utique vera narrarent, si animo illa non cernerent: quae si nulla essent, cerni omnino non possent. sunt ergo et futura et praeterita. . . . quamquam praeterita cum vera narrantur, ex memoria proferuntur non res ipsae, quae praeterierunt, sed verba concepta ex imaginibus earum, quae in animo velut vestigia per sensus praetereundo fixerunt. pueritia quippe mea, quae iam non est, in tempore praeterito est, quod iam non est; imaginem vero eius, cum eam recolo et narro, in praesenti tempore intueor, quia est adhuc in memoria mea. . . . quid cum metimur silentia, et dicimus illud silentium tantum tenuisse temporis, quantum illa vox tenuit, nonne cogitationem tendimus ad mensuram vocis, quasi sonaret, ut aliquid de intervallis silentiorum in spatio temporis renuntiare possimus? . . . nam quod eius iam peractum est, utique sonuit, quod autem restat, sonabit, atque ita peragitur, dum praesens intentio futurum in praeteritum traicit, deminutione futuri crescente praeterito, donec consumptione futuri sit totum praeteritum. . . . neque longum praeteritum tempus, quod non est, sed longum praeteritum longa memoria praeteriti est.

(And so for those that relate the things past, verily they could not relate true stories, if in their mind they did not discern them: which if they were none, could no way be discerned. There are therefore both things past and to come. . . . Although as for things past, whenever true stories are related, out of the memory are drawn not the things themselves which are past, but such words as being conceived by the images of these things, they, in their passing through our senses, have, as their footsteps, left imprinted in our minds. For example, mine own childhood, which at this instant is not, yet in the time past is, which time at this instant is not: but as for the image of it, when I call that to mind, and tell of it, I do even in the present behold it, because it is still in my memory. . . . But what when we measure silence: and say that this silence hath held as long time as that voice did; do we not then lengthen out our thoughts to the measure of a voice, even as if it now sounded, that so we may be able to say something of the vacant intervals of silence in a space of time? . . . for so much of it as is finished, hath sounded already, and the rest will sound, and thus passeth it on, until the present attention conveys over the future into the past: by the diminution of the future, the past gaining increase; even until by the wasting away of the future, all grows into the past. . . . Nor is the time past a long time, for it is not; but a long past time is merely a long memory of the past time.)[37]

Augustine understands his history to be an act of mind, an imposition of structure on what is not, a manipulation of images by words. But this is no solipsistic act. At its foundation and zenith is the immanence of God, in whose mind is the world and the past, as it is in Augustine's. The most fundamental of truths is the inspired word, the *logos* or *verbum* by which the world was made, which is the name given to the incarnate Christ. We still reserve the English derivative of *verbum* to designate the intersection of language and power, the part of speech that disturbs matter, which *suffers* the *action* of the verb. So strong was the identification of the word with material creation in the ancient world that neither in Judaism, nor in Christianity, nor in Hellenistic paganism is there any emanation from God that is not conceived of either as the word itself or as the result of verbal fiat. From God comes first language, and from language comes the phenomenal world. Augustine has no faith in some actuality that lies somehow beyond or behind the texts about the past. There is no past to be discovered, unless it exists in memory, words, or the mind of God. Augustine's task as a historian, then, is to create history as verbal-symbolic structure, which is also to perceive history in the texts of history, especially in the Scriptures, and to explicate the verbal-symbolic structures that God has encoded within the universe and within the human mind. To arrive at some perception, however imperfect, of the immanent mind of God is to arrive at the truth of history.

Augustine's verbal conception of time and duration has profound effects on his idea of the sixth millennium, the present time. It is a millennium, a number of completion, and therefore in one sense long; but it is also a symbolic number, not necessarily literal, and its duration is illusory. Its very unitary completeness gives it a sense of stasis: First and Second Comings seem infinitely withdrawn into past and

future, and the twin agonies of belatedness and expectation are alleviated. And yet both advents are made near by the very uniformity of
this period. Notice that although there is a progression of revelation
that marks the boundaries between the millennia, within each millennium is essential uniformity, and this present age is the last before the
end. There will be nothing new in time, and by this stasis all—divine
events and historical events—is made simultaneously far and near.
History, by a process of symbolization, has been removed from historical experience, which is, after all, the goal of any system of history:
to be the last. In Augustine's scheme, the world has become what it is;
the narrative of preparation has reached repletion, and can now be
contemplated in its completeness. It is therefore a retrospective narrative, without movement. The book once read becomes a simultaneous artifact in the reader's memory, the temporal process of reading,
of unveiling, having been overcome. So it is with Augustine's past,
which has the closure and finality and finitude of a text bound between
two covers of Creation and *telos* as coherently as Jerome's Vulgate:
the incoherent, unknowable past becomes a walled city of the mind.

There are places in this structure where strain is visible: like Eusebius, Augustine reveres the martyrs, whose extreme faithfulness unto
death validates the faith of the many, provides the intersection point
between faith as a verbal formula and faith as a living sacrifice. But
the age of martyrs is past, and Augustine therefore reveres the relics of
martyrs. So developed was the practice by Augustine's time, with
established shrines and offerings, that he takes considerable pains to
distinguish the practice from pagan worship:

> Quicumque etiam epulas suas eo deferunt—quod quidem a Christianis
> melioribus non fit, et in plerisque terrarum nulla talis est consuetudo—,
> tamen quicumque id faciunt, quas cum apposuerint, orant et auferunt, ut
> vescantur vel ex eis etiam idigentibus largiantur, sanctificari sibi eas volunt
> per merita martyrum in nomine domini martyrum. Non autem esse ista
> sacrificia martyrum novit qui novit unum, quod etiam illic offertur, sacri-
> ficium Christianorum.

> (Some even bring their food to the shrines—this is not done by Christians
> of the better sort, and in most countries the custom is unknown—but in
> any case those who do so say a prayer when they have laid the food down
> by the shrine, and then take it away to eat it or to bestow some of it also
> upon the needy. Their desire is that the food should be made holy for them

through the merits of the martyrs in the name of the Lord of the martyrs. But that this is no sacrifice offered to the martyrs is well known to anybody who knows the one sacrifice of the Christians, which is offered there as well as elsewhere.)[38]

Although Ambrose had condemned the practice of offering food at the martyrs' shrines, Augustine supports it with reservations. He wholeheartedly endorses other forms of veneration of martyrs, and believes in the efficacy of relics to work miracles.[39] Relics are for Augustine one of the modes of union with the past, one of the invocations of that lost time when God spoke directly. But once again the very need for a verbal doctrine of presence, for invocation, invokes also the absence that necessitates it. The doctrine of relics (which was based, for the Middle Ages, essentially on Augustine's approval) is designed (on an unconscious level—to admit the human origin of any doctrine is to invalidate it) to declare the uniformity of time, the immanence of the past within the present. What it indicates in conjunction with the process I have traced in Eusebius, however, is a further shift, a further removal, in a devolutionary chain. From Christ himself to the apostles who had known him, to the elders who had known the apostles, to the words and deeds of Christ and the apostles represented in a series of (at first disputed) texts, to the single, coherent and accepted text of the Vulgate, to the martyrs whose agony validated present faith, to the relics of those martyrs—these are the accumulated objects of faith, each new one being a medium through which its antecedents can be perceived. While each step in this chain attempts to unite time with time, present with past, the series shows only movement away from. It contains not only immanence, but also, more persuasively, otherness, the difference of the past. But the conscious doctrine must deny the series, must deny the historical and temporal nature of its own foundations, must declare each neologism to be original; if the series is admitted, it becomes a (potentially) infinite regression of simulacra, pale shadows becoming paler.

Moving diachronically through the six millennia of history are the two cities, the earthly and the heavenly, the city of men and the city of God. As the pattern of the six millenia unites past, present, and future, and makes habitable for the people of God the absence between advents, so the figure of the two cities unites Christ and Rome, and

provides a mode of being for the Church in the world. In contrast to
Tertullian's "Adeo quid simile philosophus et Christianus? Graeciae
discipulus et caeli? (But then what have Christian and philosopher in
common,—the disciple of Greece and the disciple of heaven?)"[40]
stands Augustine's "utimur et nos pace Babylonis (we also profit from
the peace of Babylon)."[41] Augustine's syncretism can be seen both in
his respectful consideration of philosophy and in the figure of the cities. One of the distinctives of Christian history from Africanus on is
that it is not a racial or tribal history; it knows as little of cultural distinctions as it does of temporal:

> Ac per hoc factum est ut, cum tot tantaeque gentes per terrarum orbem
> diversis ritibus moribusque viventes multiplici linguarum armorum ves-
> tium sint varietate distinctae, non tamen amplius quam duo quaedam gen-
> era humanae societatis existerent, quas civitates duas secundum scripturas
> nostras merito appellare possemus. Una quippe est hominum secundum
> carnem, altera secundum spiritum vivere in sui cuiusque generis pace vol-
> entium et, cum id quod expetunt adsequuntur, in sui cuiusque generis pace
> viventium.

> (And thus, in consequence, notwithstanding the many great nations that
> live throughout the world with different religious and moral practices and
> are distinguished by a rich variety of languages, arms and dress, neverthe-
> less there have arisen no more than two classes, as it were, of human soci-
> ety. Following our Scriptures, we may well speak of them as two cities. For
> there is one city of men who choose to live carnally, and another of those
> who choose to live spiritually, each aiming at its own kind of peace, and
> when they achieve their respective purposes, they live such lives, each in its
> own kind of peace.)[42]

By city, Augustine means "a gathering of rational beings united in fellowship by a common agreement about the objects of its love (ration-
alis multitudinis coetus, rerum quas diligit concordi communione
sociatus)."[43] It is ever the case with Augustine that where one might
expect to find a simple contrast there is instead an exquisite complexity. The city of men also seeks peace, is a fellowship of love:

> Fecerunt itaque civitates duas amores duo, terrenam scilicet amor sui
> usque ad contemptum Dei, caelestem vero amor Dei usque ad contemp-
> tum sui. . . . Terrena porro civitas, quae sempiterna non erit (neque enim,
> cum extremo supplicio damnata fuerit, iam civitas erit), hic habet bonum

suum, cuius societate laetatur qualis esse de talibus laetitia rebus potest. Et quoniam non est tale bonum ut nullas angustias faciat amatoribus suis, ideo civitas ista adversus se ipsam plerumque dividitur. . . . Non autem recte dicitur ea bona non esse quae concupiscit haec civitas, quando est et ipsa in suo humano genere melior. . . . Miser igitur populus ab isto alienatus Deo. Diligit tamen etiam ipse quandam pacem suam non inprobandam, quam quidem non habebit in fine, quia non ea bene utitur ante finem. Hanc autem ut interim habeat in hac vita, etiam nostri interest, quoniam, quamdiu permixtae sunt ambae civitates, utimur et nos pace Babylonis; ex qua ita per fidem populus Dei liberatur ut apud hanc interim peregrinetur.

(The two cities then were created by two kinds of love: the earthly city by a love of self carried even to the point of contempt for God, the heavenly city by a love of God carried even to the point of contempt for self. . . . Now the earthly city will not be everlasting, for when it is condemned to final punishment, it will no longer be a city. It has its food here on earth and rejoices to partake of it with the sort of joy that can be derived from things of this sort. And since this good is not of the sort to cause no difficulties for those who love it, the earthly city is generally divided against itself. . . . It is incorrect, however, to say that the goods that this city covets are not good, since through them even the city itself is better after its own human fashion. . . . Wretched, therefore, is the people that is alienated from that God. Yet even this people loves a peace of its own, which must not be rejected; but it will not possess it in the end, because it does not make good use of it before the end. But that it should possess this peace meanwhile in this life is important for us, too, since so long as the two cities are intermingled we also profit from the peace of Babylon; and the people of God is by faith so freed from it as meanwhile to be but strangers passing through.)[44]

Here is no exhortation to *anachoresis*, the condition of anchorites, literally displacement, being outside and against the inhabited place. For Augustine, the very motive and form of the city—all those qualities of the human race that necessitate the establishment of community—are necessarily good. The city of God began with the creation of the angels; the city of men began with the city founded by Cain. Founded by a fratricide as it was, the condition of that city was yet an imitation of the city of God. By typological participation, each city of men embodies the one earthly city, and is an embodiment, pale and perverse though it be, of that one city whose founder is God. The two cities are not opposites, but form and emanation; and as the earthly city cannot be conceived without the grace that is the heavenly city, so the heavenly city, invisible in this present age and world, cannot be conceived of without the analogy of its earthly counterpart:

> Pars enim quaedam terrenae civitatis imago caelestis civitatis effecta est, non se significando, sed alteram, et ideo serviens. Non enim propter se ipsam sed propter aliam significandam est instituta, et, praecedente alia significatione, et ipsa praefigurans praefigurata est. . . . Invenimus ergo in terrena civitate duas formas, unam suam praesentiam demonstrantem, alteram caelesti civitati significandae sua praesentia servientem.

> (A certain part of the earthly city has been used to make an image of the heavenly city, and since it thus symbolizes not itself but the other, it is in servitude. For it was established not for its own sake but to symbolize another city, and since it too was anticipated by another symbol, the foreshadowing image itself was also foreshadowed. . . . We find then in the earthly city two aspects: in one it manifests its own presence and in the other it serves by its presence to point to the heavenly city.)[45]

The earthly city, then, is a manifestation of grace, although it will not endure. It is to be condemned, but also loved and pitied. It contains evil, but is in its essence good. Even perverse love is yet love, and love is what creates the city. The Christian is not to reject the earthly city, but to be a citizen of that part of it which images the heavenly city, realizing, though, that the earthly city will not endure, and counting his true citizenship to be of that invisible heavenly city, implicit in the earthly, which will become explicit at the end of the age.

In the pre-Constantinian empire, martyrdom, faithfulness-unto-death, was the Church's only mode of being in the world. This utter marginality was imposed from without by the world's hatred. But in a Christianized empire, a new imaginative engagement with the earthly life became necessary. This Augustine provided by his figure of the two cities, a figure that has dominated Catholic ecclesiology to the present. The figure is superbly paradoxical and compassionate, both spiritual and worldly, allowing the Church to be both a sojourner in the land and a force in the secular city, seeking always the good of the city, the elusive, fragile intersection point at which the images of the two cities interpenetrate and become one. Combined with the six millennia, it gives the Church a place within history and time, resolving the agony of apocalyptic expectation in a redemption of the present time and the earthly city.

Augustine's new history demanded a new eschatology, as all historical revolutions must (Foucault's *L'Archéologie du savoir* and *Les Mots et les choses* are, for instance, eschatological in their anticipation of a future without the human subject). Augustine dismisses chiliasm, the belief that the millennium of the twentieth chapter of Revelation will be a literal reign of Christ and his saints on earth, following his return, as a construct of absurd fables ("ridiculas fabulas"), believable only by the carnally minded ("carnalibus").[46] Instead, he identifies this millennium with the sixth of his own scheme, the age of the Church, and it is in this age that Satan is bound.[47] The beast[48] Augustine does not identify with the antichrist, whom he believes is yet to come,[49] but with the irreligious city ("impia civitas"),[50] that part of the earthly city that does not point to the city of God. The image of the beast[51] he identifies as "those who profess belief but live as unbelievers (qui velut fidem profitentur et infideliter vivunt)."[52]

A reading of the relevant chapters of Revelation will indicate that something very peculiar is taking place in Augustine's interpretation. Revelation belongs to, or is at least about, one of the early periods of persecution, either the Neronian or the Domitianic. The great harlot, drunk with the blood of the martyrs (17:6) is a great city (17:18) that sits on seven mountains (17:9) and is obviously to be identified with Rome. Although the precise connection of the harlot/city with the beast is obscure, the beast seems to be one of the seven kings (17:10–11). This, coupled with the passage about the beast's image (13:11–18), would seem to indicate that both passages have something to do with the imperial cult, which was the direct cause and motive of all of the officially directed persecutions, and that the image (εἰκών, translated by Jerome, and hence Augustine, as the vaguer *imago*) is an idol or statue, such as the likeness of himself Caligula sought to have erected, the ambition being curtailed by his assassination, in the temple at Jerusalem.

Although the symbolism (if it is symbolism in any modern sense) is so impenetrably obscure as to be indeterminable in any systematic way, these are the obvious historical associations, those most likely to have been made by the book's contemporary audience. Augustine positively flees from any Roman and imperial identifications, even to the extent of giving a very tenuous and even distorted reading of *eikon/*

imago. What had relevance for a persecuted Church no longer has relevance for a Roman Church, the accepted religion of the empire. Such ideas are not only useless to but also subversive to Augustine's syncretism, a syncretism that made the Church at home both in time and in the Roman world, as well as in heaven; that must consider, if only to refute, all pagan philosophies as well as Christian doctrine; that must account for the pseudo-Sibylline oracles and Pliny's monstrous races in its history, just as it must account for the Scriptures. Augustine's history brooks no exclusions. It attempts the unity of secular and sacred, past and present into a single complex paradoxical union. As Peter Brown has noted, the more ascetically spiritual the Church became, the more secular power it appropriated.[53]

Augustine entrusted to a protegé, Paulus Orosius, the task of expanding on the apologetic purpose of the third book of *De civitate Dei*: to show that Christianity, and the consequent abandonment of the old gods, was not responsible for the Gothic invasions and other misfortunes. The result was his *Historiarum adversum paganos*.[54] As a source of chronology and story, Orosius is extremely important to medieval historiography (although the nature of that influence is problematic);[55] as a formulator of historical theory, a builder of the shape of time, he is an interesting but minor footnote to Augustine. Breisach gives the impression that Orosius initiates a major break with Augustinian history, that he identifies Rome with that part of the earthly city which images the heavenly (in contrast to Eusebius's supposedly critical stance toward Rome), and opposes his scheme of the four empires to Augustine's six millennia.[56] In fact, Orosius remains almost slavishly true to Augustine's commission, addressing him as his master in the dedication. His attitude toward Rome is highly critical;[57] not only does he detail the strife and crimes of the pagan empire, but he even criticizes Constantine for his persecution of Arians and pagans.[58] For Eusebius to have done this would be unimaginable; indeed it is Eusebius, not Orosius, who comes nearest, dangerously near, to identifying Christianity with a Christianized empire. The pattern of the four empires, Assyro-Babylonian, Macedonian, Carthaginian, and Roman—possibly, but nowhere explicitly, derived from Daniel's idol vision[59]—is complementary, rather than opposed, to Augustine's six millennia, and is consistent with both Augustine's and Orosius's apologetic purposes. Orosius intends to prove that Christianity is not responsible for Rome's present misery; his method is to chronicle the crimes and disasters of Rome's past, in which its pagan worship did not prevent, and even contributed to, untold human suffering, caused by the city's inherent addiction to strife and power. His is a history of the darkest aspects of the social contract, concerned with a secular world, not with the dispensations of salvation, in which the four empires become the organizing forces in a history uniform with brutality.

Orosius does contribute two new ideas that enforce the tendencies of Christian historiography toward the denial of mutability: first, his pattern of four empires, like that of the six millennia, places the present in the last age: the Roman Empire, like the millennium of the Church, will remain until the end. In neither scheme is there to be any change; both deny the possibilities of future innovation or mutation. The second of these ideas has even more profound consequences for the medieval mind: Orosius minimizes the importance of Alaric's sack of Rome and of all the other Gothic incursions. For him, the miseries of all peoples in all ages are essentially uniform, necessary concomitants of existence in a fallen world. He argues against any decline in Roman power, thus making straight a path for the medieval belief that the Roman Empire had never fallen, but had continued unchanged in its Holy Roman mimic.

With Orosius, the model of history that would dominate the Middle Ages is essentially complete, and without an understanding of this initial Christian historiographical structure, later modes of Western historical consciousness can only be miscomprehended. The model's motive is a horror of absence, of separations, of otherness between time and time—perhaps corresponding imaginatively to the *horror vacui* of medieval art and architecture—determined by the peculiar psychohistorical burden of a historically, rather than philosophically or mythically, based faith; the model's structure and method is a verbal, and hence mystical, evocation of the presence of the past through the denial of mutability and change, effectively a denial of time. This is a history, an intellectual city, in which everything is present, included within its walls. It is history as a sacrament, a transubstantial act of evocation of the intangible, the past, made tangible to the feeling intellect of faith. The created city, the cultivated garden, sought to encompass everything within its walls by a process of limitation; and

Barry University Library
Miami, FL 33161

the very presence of those walls implied an unacknowledged disjunc-
ture, a something without.

So far, I have been concerned with how and why the city of the
medieval past was built; next, I will examine its finished and derived
form, the circumscription of its architecture, the espalier of its orna-
mental trees.

2

The Completed Past

> Nothing dies; there are no empty
> Spaces in the cleanest-reaped fields.
>
> —PATRICK KAVANAGH, "On Looking into E. V. Rieu's Homer"

The Canons of Lyon in 1140 instituted a new religious festival, a feast of the Conception of the Blessed Virgin, and the action provoked a letter of rebuke from Bernard of Clairvaux. Certainly Bernard had theological and logical objections to the doctrine of the Immaculate Conception, but he does not put these first in his letter, nor does he give them precedence of emphasis. Here, commencing at the very beginning of the letter, is his principal objection:

> Inter Ecclesias Galliae constat profecto Lugdunensem hactenus praeeminuisse, sicut dignitate sedis, sic honestis studiis, et laudabilibus institutis. Ubi etenim aeque viguit disciplinae censura, morum gravitas, maturitas consiliorum, auctoritatis pondus, antiquitatis insigne? Praesertim in officiis ecclesiasticis haud facile unquam repentinis visa est novitatibus acquiescere, nec se aliquando juvenili passa est decolorari levitate Ecclesia plena judicii. Unde miramur satis, quid visum fuerit hoc tempore quibusdam vestrum voluisse mutare colorem optimum, novam inducendo celebritatem, quam ritus Ecclesiae nescit, non probat ratio, non commendat antiqua traditio. Numquid Patribus doctiores, aut devotiores sumus? Periculose praesumimus, quidquid ipsorum in talibus prudentia praeterivit. Nec vero id tale est, quod nisi praetereundum fuerit, Patrum quiverit omnino diligentiam praeterisse.

(Among all the churches of France the church of Lyons is well known to
be pre-eminent for its dignity, sound learning, and praise-worthy customs.
Where was there ever so flourishing strict discipline, grave conduct, ripe
counsels, and such an imposing weight of authority and tradition? Especially in the offices of the Church, has this church, so full of judgement,
appeared cautious in adopting novelties, and careful never to permit its rep-
utation to be sullied by any childish levity. Because of this I marvel exceed-
ingly that some of you should wish to tarnish the lustre of your good name
by introducing at this time a new festival, a rite of which the Church knows
nothing, of which reason cannot approve, and for which there is no authority in tradition. Are we more learned or more devout than the Fathers? To
introduce something which they, with all their prudence in such matters,
passed over in silence, is a most dangerous presumption. It is not as if they
would have passed it over without good reasons, for it is a thing that could
not have escaped their attention.)[1]

At the conclusion of the same letter, Bernard writes of "novitas,
mater temeritatis, soror superstitionis, filia levitatis (novelty, the
mother of rashness, the sister of superstition, the daughter of levity)."[2]

In a letter of a few years later, to Pope Innocent II, urging that the
pope take action against Peter Abelard and his followers and ban his
books, Bernard complains of Abelard's heresy in these terms:

Novum cuditur populis et gentibus Evangelium, nova proponitur fides,
fundamentum aliud ponitur praeter id quod positum est. De virtutibus et
vitiis non moraliter, de Sacramentis Ecclesiae non fideliter, de arcano sanc-
tae Trinitatis non simpliciter nec sobrie disputatur: sed cuncta nobis in
perversum, cuncta praeter solitum, et praeterquam accepimus, ministran-
tur.

(A new gospel is being forged for peoples and for nations, a new faith is
being propounded, and a new foundation is being laid besides that which
has been laid. Virtues and vices are being discussed immorally, the sacraments of the Church falsely, the mystery of the Holy Trinity neither simply
nor soberly. Everything is put perversely, everything quite differently, and
beyond what we have been accustomed to hear.)[3]

In both of these examples, the primary appeal is not to a sense of
logic, nor to a source of authoritative doctrine, neither the Bible nor
the Church (although these are appealed to in both letters). Bernard's
primary argument addresses itself to a sense of time, to a presumably
shared abhorrence of "novelty, the mother of rashness, the sister of

superstition, the daughter of levity." So powerful a hold has this sense of novelty on his consciousness and on the consciousness he attributes to his readers, that he makes no attempt to argue that novelty is wrong; it is sufficient merely to demonstrate that a doctrine or practice is new for it to be wrong as a necessary consequence, one that need not even be stated. In the letter to the Canons of Lyon, the doctrine and festival of the Immaculate Conception are wrong simply because they are not sanctioned by the past, "the Fathers."

Bernard's argument here must be distinguished from the later Protestant position that authoritative revelation had ended with the apostles. Bernard believes that the apostolic authority extends into his present, and by "fathers" he means all of the great texts of Christian tradition at least until the time of Jerome and Augustine, if not beyond. His argument is for a completed past, a completed body of doctrine and a completed knowledge of the world: "If this is true, why was it not known to the past?" is in effect what he writes. In the letter to Pope Innocent, he damns Abelard with "a new gospel . . . a new faith . . . and . . . a new foundation." This newness he then equates with perversity and difference, and equates these in turn with anything "beyond what we have been accustomed to hear."

Bernard holds a unified and completed view of knowledge, which demands as a desideratum a unified and completed history. Abelard's *Sic et non*, a book of statements from the Bible and the "Fathers" placed in contradictory juxtaposition to show that to every authoritative proposition there is an authoritative denial—and this with no attempt at resolution—was, until Wycliff, culturally the most dangerous, the most antimedieval of all medieval books. It was not in the old Gnosticism of the Cathars that the delicate medieval mental world would find its destruction, but in the new nominalism of Abelard, the connoisseurship of incoherence. Yet even if Abelard's doctrine had not included epistemological incoherence, he would still, in Bernard's world, be guilty of temporal incoherence like the Canons of Lyon. Abelard's first sin was that he had said something new.

The idea of history that lies behind Bernard's statements is one in which Foucault's idea of reconstitution as the goal of total history is irrelevant: to Bernard the past needs no reconstitution because it is already constituted. Past and present exist as a single static whole, a unity, a millennium. There can be no intellectual separation from the past because the past is the arbiter of intellect. It is possible, even, that

in Bernard's world the present has a sense of unreality, of distance. It
cannot rival the demands of that great icon, that object of meditation,
the past, the place where God is to be found. Meditation on symbolic
structures—the book between two covers, the millennium between
two acts of God—draws the mind's devotion from the phenomenal,
imbuing the present world both with insubstantiality and with hallu-
cinatory brightness. By long meditation on the rhetoric of union, the
psychological experience of mutation in history is overcome. The history that informs Bernard's two letters is the triumph of that union
with the past which Eusebius had sought.

A survival of precisely the same mode of historical thought can be
found as late as 1521 in Henry VIII's *Assertio septem sacramentorum*,
in the light of subsequent events a surprising and ironic exemplar:

> Quae pestis unque tam perniciosa invasit greges CHRISTI? Quis serpens
> unque tam venenatus irrepsit, quis, qui de babylonica captivitate ecclesie
> scripsit: qui scripturas sacras ex suo sensu contra CHRISTI sacramenta detor-
> quet: traditos ab antiquis patribus ecclesiasticos ritus eludit: sanctissimos
> viros, vetustissimos sacrarum literarum interpretes, nisi quarenus ipsius
> sensui conveniunt et consentiunt, nihilipendit: sacrosanctam sedem
> romanam Babylonem appellat: summum pontificium vocat tyrannidem:
> totius ecclesiae decreta saluberrima captivitatem censet: sanctissimi ponti-
> ficis nomen in Antichristum convertit. . . . Mirum est igitur, ex tot sanctis
> patribus, ex tot oculis, quot, in ecclesia tam multis saeculis, idem legerunt
> evangelium, nullumfuisse unque tam perspicacem, ut rem tam apertam
> deprehenderet . . . nemo est, opinor, qui credet hac in parte Luthero, ni si
> primum doceat aut aliud evangelium legisse se, qui sancti illi patres lege-
> runt, aut illud idem, vel legisse diligentius: vel intellexisse melius, aut sibi
> denique ma iorem esse curam fidei, qui ulli unque hactenus mortalium
> fuerit.

> (What *Plague* so Pernicious did ever invade the *Flock of Christ*? What *Ser-
> pent* so Venomous has crept in, as he who writ of the *Babylonian Captivity
> of the Church*? Who wrests Holy *Scripture* by his own Sense against the
> *Sacraments of Christ*, and abolishes the *Ecclesiastical Rites and Ceremo-
> nies* left by the *Fathers*, undervalues the most Holy and Antient Interpret-
> ers of Scripture, unless they concur with his Sentiments; calls the most Holy
> See of *Rome, Babilon*, and the *Pope's* Authority, *Tyranny*: and Esteems the
> most wholesom Decrees of the *Universal Church* to be *Captivity*; and turns
> the Name of the most Holy Bishop of *Rome*, to that of *Antichrist*. . . . 'Tis
> a wonder that of so many *Holy Fathers*, of so many Eyes which have read
> the *Gospel* in the *Church* so for many Ages [*sic*], that none was ever so

quick-sighted, to as [*sic*] perceive a thing so apparent. . . . I suppose there
are none will believe him, unless he first shew that he has Read another
Gospel different from that the *Holy Fathers* ever Read, or that in Reading
the same he has been more diligent then They, or has better understood it;
or finally, that he is more careful about *Faith*, than ever any man before
him was.)[4]

The essence of Henry's argument is the impossibility of a true novelty: why was not Luther's interpretation known to the "*Holy
Fathers*," the "Antient Interpreters of Scripture"? As did Bernard,
Henry accuses his neoteric opponent of "another *Gospel* different
from that the *Holy Fathers* ever Read," but instead of condemning
him for it, ludicrously challenges him to produce it. Once again the
doctrine cannot be discussed as doctrine; it is invalidated by the mere
fact that it has not existed before. Henry also makes explicit what is
implicit in Bernard: in the unity and uniformity of time within the
sixth millennium, the unity and uniformity of the Catholic Church
inheres. He locates the Church primarily in the past, not the present.
Its unity throughout time is essential to its unity in the present. If the
present Church is severed from the past Church, Henry knows, then
each limb of the present Church must be severed from each. This identity of temporal and present geographical fragmentation is one of the
major concerns of the *Assertio*:

Alioqui si perstet in eo Lutherus, ut ecclesiam papae discernat ab ecclesia
Christi, et apud alteram, dicat ordinem haberi pro sacramento, non haberi
apud alteram, proferat illam ecclesiam Christi, quae contra fidem papalis
(ut vocat) ecclesiae, ignorat sacramentum ordinis. Interim certe perspi-
cuum est, quum dicat hoc sacramentum ignorari ab ecclesia Christi, et de
Christi ecclesia, dicat eos, quibus praesidet papa, non esse: utraque ratione
ab ecclesia Christi eum segregare non Romam tantum, sed Italiam totam,
Germaniam, Hispanias, Gallias, Britannias, reliquasque gen tes omnes,
quaecunque romano pontifici parent, aut or dinem pro sacramento reci-
piunt. Quos populos omnes quum de Christi tollat ecclesia, necesse est ut
aut ecclesiam Christi fateatur esse nusque, aut more donatistarum, ecclesiam Christi catholicam, ad duos aut tres haereticos redigat de Christo
susurrantes in angulo.

(Otherwise, if Luther persists in his distinction of the *Popes Church*, from
Christs; and in saying that the one has *Orders* for a *Sacrament*, the other
not. Let him shew us the *Church* of *Christ*, which, contrary to the *Faith* of
the *Papal Church*, (as he calls it) knows not the *Sacrament* of *Order*. In the

mean while it appears evidently, that by asserting this *Sacrament* to be unknown to the *Church* of *Christ*, and that they are not of *Christs Church* who are govern'd by the *Pope*: He seperates, by both these Reasons, from *Christ*'s *Church*, not only *Rome*, but also all *Italy*, *Germany*, *Spain*, *France*, *Britain*, and all other Nations, which obey the See of *Rome*; or have *Orders* for a *Sacrament*. Which People being by him taken from the *Church* of *Christ*; it consequently follows, that he must either Confess *Christs Church* to be in no place at all, or else, like the *Donatists*, he must reduce the *Catholick Church* to two or three *Hereticks* whispering in a Corner.)[5]

Both of these examples, Bernard of Clairvaux and Henry VIII, exemplify the medieval structure of time in its particular psychological and doctrinal manifestations. The emphasis of both is on unity, a temporal unity, a communion with the past without which political, philosophical, and spiritual unity are seen as impossible. The perception of this unity of time is, in both cases, so extreme that no doctrine can be considered on its logical or biblical merits, that is, considered atemporally, or, within the medieval understanding of history, ahistorically. All that is true must exist within the single fabric of a history that will not allow for development or for the new, for secrets hidden within time and reserved only for latter generations, because the idea of development and change enfolds within itself the incipient contingency, as inevitable in its eventual geometry as the growth of crystals, of the unknowability of God. As with the development of the rhetoric of union at the inception of Christian history, so with its derivative, realized form: the motivation for syncretic unity is the repressed vision of chaos, of a universe so dynamic that no agreement, no surety, no lasting knowledge, no standing place can be found. Both Bernard and Henry address their rhetorics of unity against an enemy—Abelard, Luther—although in Henry's case, only thirteen years later, the enemy would prove to have been himself.

Before I proceed to an examination of the various manifestations and mediations of the doctrine of temporal unity, the underlying, denied substrata of chaos deserve a brief exhumation. Within almost all medieval histories, a tremendous tension between the transcendent and immanent realms is perceivable. In the universal scheme of time there is harmony and form, but in the chronicling of human events within that scheme, the medieval historian reveals an extraordinary toleration of chaos. An excellent exemplar is Dante's *Commedia*, in which the affirmations of the universal order, the perfectly proportioned architectonics of hell, purgatory, the spheres, and the poem itself, contrast almost infinitely with the squalid, inchoate human history revealed in the stories of the shades. In Dante the human chaos does not, as would be expected in a modern (in the loosest sense) work, subvert the divine order, but instead the opposite is true: by its contrast with the temporal, the divine order is exalted.

The chaos of the human realm impresses itself cumulatively upon the reader of medieval histories, but it can be sampled in small in these first lines of Gregory of Tours's sixth-century *Historiae Francorum*:

> ... cum nonnullae res gererentur vel recte vel inprobae, ac feretas gentium desaeviret, regum furor acueretur, aeclesiae inpugnarentur ab hereticis, a catholicis tegerentur, ferveret Christi fides in plurimis, tepisceret in nonnullis, ipsae quoque aeclesiae vel ditarentur a devotis vel nu darentur a perfidis.

> (A great many things keep happening, some of them good, some of them bad. The inhabitants of different countries keep quarrelling fiercely with each other and kings go on losing their temper in the most furious way. Our churches are attacked by the heretics and then protected by the Catholics; the faith of Christ burns bright in many men, but it remains lukewarm in others; no sooner are the church-buildings endowed by the faithful than they are stripped bare again by those who have no faith.)[6]

Here is a formula for a history without plot or narrative, without development or *telos*. In the divine plan for the redemption of the world there may be an architectonic unfolding, but in the human realm there is only the monotony of chaos in the illusory extension of time between the First and Second Advents. R. G. Collingwood attributes this sense of chaos (at least as experienced by modern readers), rightly I think, to an absolute opposition in the medieval mind between the objective purpose of God and the subjective purposes of man, so extreme that it makes human history almost irrelevant, except in those places where it touches the purpose of salvation, and this only to the salvation or damnation of individual actors within history.[7] Although there are missionaries and saints and just kings in medieval histories, the overriding principle, according to Collingwood, is that God does not need human agents. This may be an overstatement, but it is a true one. Even in the histories of the Crusades—which surely deal with human protagonists acting as divine agents, which surely ought to have a narrative structure, a plot, a thesis—the overwhelming theme is of the failure of human action to accomplish the divine will. The mission always degenerates, from its inception, into the confusion of intestine wars among the Christians, the sack of Christian Constantinople by its supposed relievers, and slaughter and greed in the name of God, and it ends in a failure to hold what has been won. In many cases even the excoriating moral is not drawn. Fulcher of Chartres, for instance, can combine the most abject slaughter and pillage—the cutting open of bowels to find swallowed bezants, the killing of women rather than raping them, this related with approval as a lesser sin— with unaffectedly sincere Christian devotion. There is for him no difference, no disjunction between cruelty and Christian devotion. All is told with a terrible acceptance that reduces all deeds to irrelevance. In many instances it is not possible to detect in him any authorial attitude.[8] Stripped of its religious motive, this shapeless tolerance will later become the chivalrous approbation of Froissart and Chastellain: histories of endless, meaningless warfare, of vainglory enthroned, with neither criticism nor irony, as the arbiter of earthly history. Orosius saw a similar chaotic strife, without development or decline, without plot or progress, in the history of Rome, but he, as did the greater and earlier medieval historians, saw something wrong with it. In later medieval histories, the objective purpose of God becomes so entirely removed from the subjective purposes of man that it disappears from the purview, no longer even a force of contrast. The effect might be imagined by envisioning Shakespeare's Wars of the Roses cycle denuded of every moral judgment: an imaginative world in which Richard III is approved every bit as much as Richmond.

The precarious tolerance of chaos within the human realm extends also to the Church. Here is Matthew Paris:

> Eodem tempore, permittente vel procurante Papa Gregorio, adeo invaluit Romanae ecclesiae insatiabilis cupiditas, confundens fasque nefasque, quod deposito rubore, velut meretrix vulgaris et effrons, omnibus venalis et exposita, usuram pro parvo, symoniam pro nullo inconvenienti reputavit, ita ut alias affines provincias, immo etiam puritatem Angliae sua contagione macularet.

> (About this time, either with the permission or by the instrumentality of Pope Gregory, the insatiable cupidity of the Roman court grew to such an

extent, confounding right with wrong, that, laying aside all modesty, like a common brazen-faced strumpet, exposed for hire to every one, it considered usury as but a trivial offence, and simony as no crime at all; so that it infected other neighbouring states, and even the purity of England, by its contagion.)[9]

This extraordinarily condemnatory rhetoric belongs not to a Calvinist but to a monk of Saint Albans Abbey, an orthodox and unschismatic Catholic, and it is written not in the sixteenth century but in the thirteenth (the events described in this passage occur in 1241).

The histories show an awareness of, and yet a tolerance of, every kind of corruption within the Church; yet, to the medieval mind, Paris's included, the shortcomings of persons did not impugn the institution (this in the sense of God's instituting, rather than the modern connotation). This seeming contradiction could be reconciled, because the Middle Ages thought of the Church not as a synchronic structure, existing only in its present state, but as a diachronic mystery present throughout the past and into the future, as unchangeable as God. So Dante in *De monarchia* does not debate the Church (there was no holding place in Dante's mind for such a debate), although he opposes its claims to secular power, but only certain misguided advocates of the Church: persons but not institutions or offices.[10] Both Dante and Henry VIII place the Church they describe almost entirely in the past, inviolable and unchallengeable.

But if human vagaries could be tolerated, the appearance of any institutional variance could not. Thus, according to the Monk of Saint Gall, Charlemagne was concerned by the ineluctable tendency of liturgy to mutate, to develop local and temporal difference:

Referendum hoc in loco videtur, quod tamen a nostri temporis hominibus difficile credatur, cum et ego ipse, qui scribo, propter nimiam dissimilitudinem nostrae et Romanae cantilenae non satis adhuc credam, nisi quod patrum veritati plus credendum est quam modernae ignaviae falsitati. Igitur indefessus divinae servitutis amator Karolus . . . sed adhuc omnes provincias immo regiones vel civitates in laudibus divinis, hoc est in cantilenae modulationibus, ab invicem dissonare perdolens.

(Here I must report something which the men of our time will find it difficult to believe; for I myself who write it can hardly believe it, so great is the difference between our method of chanting and that of the Roman, were it not that we must trust rather the accuracy of our fathers than the false suggestions of modern sloth. Well then, Charles, that never-wearied lover of the service of God . . . was grieved to observe how widely the different provinces—nay, not the provinces only but districts and cities—differed in the praise of God, that is to say in their method of chanting.)[11]

Charles asked Pope Stephen, so the story continues, for twelve clerks who would teach the various Frankish churches the Roman rite, but these clerks, motivated by envy when they saw the splendor of the Franks, conspired among themselves to teach their host churches different, novel liturgies, and thus compounded the problem. They were sent back to the pope, by this time Leo, who condemned them to perpetual imprisonment. The problem was apparently rectified by sending Frankish clerks to Rome for instruction, but, as is evident from the first sentence of the quotation above, the liturgy had changed anew by the time of writing.

Here is a problem that touches the efficacy of the Church in a way that simony or priestly concubinage never could. It is a difference in the institution of the Church itself, in the form of worship, and can touch not only that tiny part of the Church that exists in the present, but can damage the whole edifice in every age. If the form of the Church is changing, mutable, then its delicately counterbalanced geometry nears the limits of endurance. The Church only has authority, only coheres, to the extent that it reaches into the past, that it unites in mystical participation Christ and the apostles to the present. If the Church itself incorporates difference—and regional difference is necessarily temporal difference, for such a mutation can only take place within time (that two regions differ from each other means that both must differ from the past, from the one original form)—then it may not be the apostolic Church, but only some degraded simulacrum thereof. Temporal unity is the essential balance (to borrow Henry Adams's image in *Mont Saint Michel and Chartres*), the flying buttress, that holds the great weight and burden of the Church in equilibrium. In the stress on unity, on uniformity, the fragility of balance is revealed (Charles's painstakingly achieved uniformity does not last into his biographer's present).

As always, the formulation of a verbal consciousness implies the subverbal experience of its opposite; the rhetoric of unity and coherence is necessitated by the great and probable fear of infinite incoherence—thus the strength of the medieval hatred of heresy, of Mohammed (called the "filthy prophet" by Matthew Paris),[12] of the schismatic Eastern Church.[13] Yet clear and absolute heresy could be cast out, could strengthen the Church by the vehemence of its opposition. Far more dangerous was the imperceptible drift of variation among the orthodox, within the city's wall. This is why Bede describes at such great length the English Saint Augustine's correspondence with Pope Gregory concerning every detail of church doctrine and practice.[14] It is why those letters—so earnestly anxious that the rite or polity of the English church might differ from the Roman, that is, from the past— were written. Thus asserts Dante:

> Item dico, quod ens, et unum, et bonum, gradatim se habent secundum quintum modum dicendi "prius." Ens enim natura praecedit unum, unum vero bonum; maxime enim ens maxime est unum, et maxime unum est maxime bonum. Et quanto aliquid a maximo ente elongatur, tanto et ab esse unum, et per consequens ab esse bonum. Propter quod in omni genere rerum illud est optimum, quod est maxime unum, ut Philosopho placet in iis, quae de simpliciter Ente. Unde fit, quod unum esse videtur esse radix eius, quod est esse bonum; et multa esse, eius quod est esse malum. Quare Pythagoras in correlationibus suis, ex parte boni ponebat unum, ex parte vero mali plura, ut patet in primo eorum, quae de simpliciter Ente. Hinc videri potest, quod peccare nihil est aliud, quam progredi ab uno spreto ad multa.

> (Likewise I affirm that being and unity and goodness exist *seriatim* according to the fifth mode of priority. Being is naturally antecedent to unity, and unity to goodness; that which has completest being has completest unity and completest goodness. And as far as anything is from completest being, just so far is it from unity and also from goodness. That in every class of objects the best is the most unified, the Philosopher maintains in his treatise *on simple Being*. From this it would seem that unity is the root of goodness, and multiplicity is the root of evil. Wherefore Pythagoras in his Correlations placed unity on the side of good and multiplicity on the side of evil, as appears in the first book *on simple Being*. We can thus see that to sin is naught else than to despise unity, and to depart therefrom to multiplicity.)[15]

These then are the poles of the medieval world in contention, unity and multiplicity. Multiplicity enters the verbal, textual consciousness only to be reviled, but it is in many ways the more fundamental of the

two forces, and the horror of it motivated an immense verbal, institutional, architectural, and artistic assertion of unity. All of these
works—the manifest, physical, and conscious emanations of the
human mind—combine to form that state of cooperative being that
Augustine defined as a city.

Thus far, the boundaries of the city; I have tried to show the medieval
aesthetic assertion of temporal unity at its most precarious, where it is
conscious of the anti-aesthetic multiplicity that threatens it: the endless forest outside the circle of firelight, outside the city wall of human
order and naming. What follows will look more closely within the city,
at the details of the aesthetic structure itself.

In order to hope to understand medieval history (and one can only
hope to understand it), it is necessary to remember that history is not
the past, any more than theology is God; history is made of language
and nothing else. Medieval history can best be understood as a species
of poetry; it is far closer to epic than to the "scientific" history of the
nineteenth century. If it is evaluated in terms of the "scientific" aesthetic of history, then it can only be misunderstood. The modern historian or editor criticizes medieval histories for their anachronism or
credulity, and praises their rare cases of skepticism about sources and,
less rare, accurate relation of events.[16] To praise the first is to praise a
positive fault; to praise the second is to reward mere pedestrianism. If
we are to estimate medieval history in the terms of its own aesthetic
discourse, then the first step is to understand that anachronism and
credulity (in the sense of faith) are its highest achievements.

The medieval aesthetic of time sought union with the past through
a series of mediating icons, images through which the absent past
could be made, in both senses of the word, present. I use icon here in
opposition to simulacrum. Simulacrum, in my understanding of its
connotation, represents that part of the neutral term "image" that has
declined from the thing it represents. Icon, as I use it, means an
enabling image that ascends toward the thing itself.

The most fundamental of these icons was the incantation of unity
and syncretism, that is, the repeated and insistent assertion of unity in
all of its aspects. The doctrine tied the unity of time to a unity of
knowledge, involving a redemption (a redemption so extreme and pervasive that it denies the need for redemption) not only of the Christian

past but also of the pagan past. Eusebius asserted that Tiberius was converted by news of Christ's death, an assertion repeated by almost every medieval historian covering the period.[17] Similarly, Otto of Freising, writing in the twelfth century, states that the Greek philosophers discerned by virtue all that could be known of God without the Incarnation of Christ, and cites for his authority Augustine's testimony that he found in Plato, "In the beginning was the Word." He also repeats Augustine's tradition that Plato was the pupil of Jeremiah, although he sees, almost with an audible sigh of regret, a problem with dating.[18] Even this caveat is not an objection to what modernity would call anachronism, but is necessitated by the syncretic chronicle tradition of Africanus, Eusebius, and Jerome, by Otto's time grown into a highly developed system. It is really an example of one kind of anachronism (this time in the positive, medieval sense), that of the syncretic chronology, taking precedence over the philosophical syncretism asserted by Augustine. Here, from the chronicle ascribed to Matthew of Westminster, is a remarkable example of the extent to which this syncretic chronology permeates medieval history:

Anno igitur ab origine mundi quinquies millesimo ducentesimo minus uno, imperii Augusti Caesaris xl ii, regni vero Herodis regis Judeae xxx, regnante in Britannia Kimbelino Tenuantii filio, Olimpiadis autem centisimae nonagesimae tertiae anno tertio, quarta indictione, epactis existentibus undecim, concurrentibus quinque, octavo kalend. Januarii, natus est Dominus noster Jesus Christus et Dei Filius, in Bethleem Judae, ut populum gentilem et Judaicum sanctificando sibi connecteret et uniret. Natus est autem Sua ordinatissima dispositione, nocte diei Dominicae, quia si tabulas compoti retro percurras, invenies hujus anni concurrentem v. regularem, Januarii iii. Quibus junctis et sublatis vii., unus remanet. Itaque kalend. Januarii in Dominica invenies quod concurrit. Nam eadem die qua dixit, "Fiat lux, et facta est lux," visitavit nos Oriens ex alto, ut illuminaret sedentes in tenebris, et dirigeret in viam pacis. Inchoata est vero, secundum quosdam, sexta aetas a nativitate Christi, secundum Apostolum, qui ait, "Cum venerit plenitudo temporis, etc."; secundum alios, a die qua baptizatus est, propter vim regenerativam datam aquis; secundum alios a passione, quia tunc aperta est porta, et inchoata est septima quiescentium gloria.

Adventum itaque suum ideo Christus usque ad sextam distulit aetatem, ut novae legis plenitudinem prolixitas temporis non auferret. Nam si ante venisset, cuncta forsitan novae legis praecepta vetustas temporis delevisset.

Proinde decebat, ut qui sexta die hominem fecerat, sexta mundi aetate eum
ad plenitudinem legis perduceret, et jam veterescentem suo novo adventu
reficeret.

(In the five thousand nine hundred and ninety-ninth year after the creation
of the world, in the forty-second year of the reign of Augustus Caesar, and
the thirtieth of the reign of Herod, Cymbeline, the son of Tennancius, being
King of Britain, and in the third year of the hundred and ninety-third
Olympiad, in the fourth indiction, in the second existing epact [*sic*], and
fifth concurrent, on the twenty-fifth day of December, our Lord Jesus
Christ, the Son of God, was born in Bethlehem of Judea, in order, by sanc-
tifying the people of the Gentiles, and the Jewish nation, to bind them
together, and unite them to himself. And he was born, according to his
most exactly ordered arrangement, on the night of the Lord's day; because,
if you reckon back in the chronological tables, you will find the fifth concurrent of this year, the third regular of January. And when they are added
together, and seven are subtracted, one remains. And so you will find that
the first of January on that year fell on a Sunday, which corresponds to my
calculation. For on the same day in which God said, "Let there be light,
and there was light," the day-star from on high visited us, in order to give
light to those who sat in darkness, and to guide them into the way of peace.
And the sixth age began, according to some people, with the Nativity of
Christ. According to the Apostle, for instance, who says, "When the fullness
of time shall have come," &c. According to others, with the day on which
he was baptized, on account of the regenerative power given to the waters.
According to others, with his Passion, because then the door was opened,
and the seventh glory of those at rest began. Therefore, Christ put off his
Advent to the sixth age, in order that the length of time might not destroy
the fullness of the new law. For if he had come before, perhaps the length
of time would have effaced all the precepts of the new law. Moreover, it
was becoming that he who had made man on the sixth day, should bring
him to the fullness of the new law in the sixth age of the world, and as the
world was now growing old, should refresh it by his new arrival.)[19]

What a web of context is here given to the birth of Christ! It is dated
firstly in the age of the world, calculated from the genealogies in Genesis, secondly in the Roman secular history (the *Flores historiarum*
peculiarly calculates the reign of Augustus to commence with the
death of Julius), thirdly within the Christian gospel tradition (the reign
of Herod), then within the "fabulous" British chronicle tradition
derived from Geoffrey of Monmouth (Cymbeline), and within the
Greek ordering of time (the Olympiad). Here the various, conflicting

pasts are wedded into one, made interpenetrative. As Christ came to "bind" and "unite" Jew and Gentile, so he also bound the several past times together and to the present.

Temporal and cultural difference is further annihilated by the typology of his coming. As the first day of Creation, the day of light, was a Sunday (the chronology was this precise), so the Incarnation brought light in a play of foreshadowing and fulfillment, a dance of interfiguration. Typology (in the belief of those who accept it) is God's symbolism, independent of interpreters. In a typological history there can be no separation of time from time; all times form a symbolic, palindromatic unity, which must be read both forward and backward, each type prefiguring, each antitype completing, both incomprehensible without the other, as in Donne's words: "We think that Paradise and Calvary, / Christ's cross, and Adam's tree, stood in one place."[20]

The sixth age referred to in the second paragraph is of course Augustine's sixth millennium, and the "seventh glory" of those who were already dead at Christ's coming is Augustine's seventh day, the sabbath without an end. Here also is the persistent motif of the completion of time. Christ postponed his coming to the sixth and last age, the fullness of time, so that length of time would not destroy the gospel and the Church; and, with this phrase, the abyss beneath the incantation obtrudes into consciousness. But as time is complete in this history (which exists for the purpose of making it complete), no new thing lurks in the future to cut the believer from the believed past.

In any literature informed and determined by this completed and syncretic view of history, two correlative phenomena ought to be observable: one is anachronism, in the modern, pejorative sense of confusing customs, costume, architecture, and beliefs appropriate to various times; the other is what I will call "auctourism," the credulous dependence on authorities, on books.

Of anachronism little need be said except to posit an explanation for it. It is universal in medieval literature and art. In the mystery plays, the torturer says to Christ as he nails him to the cross:

> In fayth, syr, sen ye called you a kyng
> you must prufe a worthy thyng
>   That falles unto the were;
> ye must Iust in tornamente;
> Bot ye sytt fast els be ye shentt,
>   Els downe I shall you bere.[21]

As Jesus calls himself a king, he must joust in the tournament and sit fast on his cross. Similarly, "Duke" Theseus's Athens in "The Knight's Tale" is a medieval duchy filled with knights and tournaments, as is the Troy of *Troilus and Criseyde*. In Robert de Boron,[22] Pontius Pilate is the Duke of Jerusalem, and his knight, Joseph of Arimathea, begs Christ's body on the basis of his oath of feudal fealty. This kind of anachronism is even more evident in the visual arts: soldiers at Christ's tomb (always a European vault tomb, never the cave tomb with rolling stone) wear fourteenth-century armor; in Flemish painting each biblical event takes place in a landscape of idealized gothic cities and castles. If this is due to ignorance, then that ignorance is certainly not simple; anachronism in medieval art and literature is inextricably bound to and part of a whole philosophy of history. The doctrine of temporal unity demands that all times be alike, and that likeness extends even to the external aspects of culture: costume and architecture. It is true that the medieval painter or carver did not know that the Romans and Jews had dressed and built differently from the men of his own day; he had no reason even to suppose that such a thing was probable or possible. But it was also an ideological statement: Christ had died in a world and time precisely like his own. By anachronism, the devotee who used the art as an icon, as a medium of religious devotion (which was the function of all medieval art) was united to the object of that devotion: a Jesus not of a distant, foreign past, but of an eternal present.

If anachronism is a necessary expression of a unified, completed, and syncretic history, then "auctourism" is equally inevitable, but even more important. It is one of the mediating icons through which the past is made present. Medieval history privileged the past, as the locus of salvation. The authority of the past resides principally in texts, particularly the texts of the New Testament and of the church fathers. But because of the radical cultural syncretism of Christian history, pagan texts are also imbued with doctrinal weight; even the pseudo-Sibylline oracles were thought to contain Christian doctrine and to be as genuinely prophetic as the Revelation of John or the Jewish prophets.[23] Devotion came to be directed not only to The Book, but to all books.[24] An example: Geoffrey of Monmouth begins his *Historia regum Brittaniae*:

Cum mecum multa & de multis sepius animo revolvens in hystoriam regum britannie inciderem. in mirum contuli quod infra mentionem quam de eis gildas. & beda luculento tractatu fecerant nichil de regibus qui ante incarnationem christi inhabitaverant.

(Oftentimes in turning over in mine own mind the many themes that might be subject matter for a book, my thoughts would fall upon the plan of writing a history of the Kings of Britain, and in my musings thereupon meseemed it a marvel that, beyond such mention as Gildas and Bede have made of them in their luminous tractates, nought could I find as concerning the kings that had dwelt in Britain before the Incarnation of Christ.)

So far Geoffrey has cited two authorities, two *auctours*, Gildas and Bede. But he is going to introduce a good deal of previously unknown material, and therefore:

Talia mihi & de talibus multociens cogitanti optulit walterus oxenefordensis archidiaconus vir in oratoria arte. atque in exoticis hystoriis eruditus quendam britannici sermonis librum vetustissimum . . . actus omnia continue & ex ordine perpulcris orationibus proponebat.

(Now, whilst I was thus thinking upon such matters, Walter, Archdeacon of Oxford, a man learned not only in the art of eloquence, but in the histories of foreign lands, offered me a certain most ancient book in the British language that did set forth the doings of them all in due succession and order . . . all told in stories of exceeding beauty.)[25]

Every history must have an authoritative, ancient source, and every medieval historian who is not chronicling events within living memory begins by naming his *auctours*; in national chronicles these are generally the founders of the national history, in Geoffrey's case Gildas and Bede; in universal histories they almost always include Eusebius, Augustine, and Orosius, and sometimes also Josephus and Varro. But what is to be made of Archdeacon Walter's mysterious book? The general opinion has been, from not long after Geoffrey's own time, that the book was a mere invention.[26] Certainly there was great scope for inventing sources in a world in which a scholar would have heard of many books that he had never actually seen. What is significant is that Geoffrey would feel the need to invent a single, ancient (he stresses this) source for what scholars believe was gleaned from a haphazard collection of sources, written and oral. But Geoffrey was after all writing a history, albeit a fraudulent and fabulous one, and histories, being "true" books about the past do need sources. A second example may

clarify the real nature of "auctourism." This is Chretien de Troyes at
the beginning of *Cligés*:

> Ceste estoire trovons escrite,
> Que conter vos vuel et retreire,
> An un des livres de l'aumeire
> Mon seignor saint Pere a Biauveiz.
> De la fu li contes estreiz,
> Don cest romanz fist Crestiiens.
> Li livres est mout anciiens,
> Qui tesmoingne l'estoire a voire;
> Por ce feit ele miauz a croire.

(This story, which I intend to relate to you, we find written in one of the
books in the library of my Lord Saint Peter at Beauvais. From there the
material was drawn of which Chretien has made this romance. The book
is very old in which the story is told, and this adds to its authority.)[27]

Here is the same appeal to an *auctour*, the same emphasis, even
stronger in this example, on the antiquity of the *auctour*, its very pastness, which gives it authority. But *Cligés* is a romance, what will later
be called a fiction, yet it equally demands a past source to redeem it
from the sin of invention. The medieval mind would not tolerate the
new, the original; every work of literature or art must be a recapitulation of the universal atemporal unity of what is already known. And
a unity of knowledge and of time implies a unity of genre. There is no
difference between Geoffrey's "history" and Chretien's "romance" in
their relation to truth. Chretien believed in the truth of the matter he
was relating; this truth was attested by the antiquity of the book in
which he found it. Devotion to the past, as embodied in the texts of
that past, applied to matters secular and trivial as well as divine and
weighty. Distinctions between "true" and "fictional" genres were not
made by the Middle Ages; Froissart began his *Chronicle* in verse and
abandoned it for prose because he found it too difficult, not because
it was an inappropriate form in which to write a true history.

This same devotion to texts and to the past can also be found in the
stated motive of medieval history. The past was the great source of
moral imitation, as here in Bede:

> Sive enim historia de bonis bona referat, ad imitandum bonum auditor
> sollicitus instigatur; seu mala commemoret de pravis, nihilominus religio-

sus ac pius auditor sive lector devitando quod noxium est ac perversum, ipse sollertius ad exsequenda ea quae bona ac Deo digna esse cognoverit, accenditur.

(For whether an history shall contain good things concerning good men, the careful hearer is thereby stirred up and provoked to follow after well-doing; or whether it shall report evil things concerning froward men, the devout and well-disposed hearer or reader none the less, by flying that is evil and noisome to his soul, is himself moved thereby more earnestly to follow after the things he knoweth to be good and acceptable to God.)[28]

Here is the same motive in Matthew of Westminster:

Sed quid contra quosdam auditores pigros dicemus, qui obtrectando dicunt, quid necesse est vitas vel mortes seu diversos hominum casus litteris mandare, prodigia caeli et terrae vel aliorum elementorum scriptis impressa perpetuare? Noverint isti, bonam vitam et mores praecedentium ad imitationem subsequentium proponi, malorum vero exempla ut non imitentur, sed ut potius vitentur, describi. . . . Non sunt igitur audiendi, qui Cronicorum libros et maxime a catholicis editos dicunt esse negligendos, per quos quicquid humanae sapientiae et saluti necessarium est, per memoriam invenire, per intelligentiam cognoscere, et per facundiam proferre studiosus valeat indagator.

(But what are we to say in reply to certain dull auditors, who, in a spirit of detraction, say, "What occasion is there to commit to writing the lives and deaths of men, and different events which happen to mankind? Why perpetuate by written records the recollection of prodigies in heaven and earth, or those which affect the other elements?" Let them know that the good lives and virtuous manners of men of old time, are recorded to serve as patterns for the imitation of subsequent ages; and that the examples of the wicked are set forth, not that they may be imitated, but that they may be shunned. . . .

Those men, then, must not be listened to, who say that books of Chronicles, and especially those published by Catholics, ought to be neglected; since by the study of them a diligent enquirer may be able to discover by his memory, to understand by his intelligence, and to set forth, with eloquence, all that is necessary for human life, and for human safety.)[29]

History, the textual past, is, for Bede and Matthew of Westminster, a source of story, of imitation (the devotional *imitatio* is included in the idea). The force of the sentiment can really only be understood by contrast to the modern use of the past. We do not seek to learn philosophy from Aquinas or Nicholas of Cusa, nor military science from Caesar or Edward III, nor the effective suppression of witchcraft from Cotton Mather or the *Malleus maleficarum*; we seek from history not example, for that requires an intellectual community with the subjects of history; instead we look for patterns of development that enable us to discount and abolish the past. George Santayana's statement, quoted to the point of cliché because it is so representative of what we believe, that, "Those who cannot remember the past are condemned to repeat it,"[30] expresses precisely the opposite view of history from the medieval. Santayana's is a remembering in order to escape, akin to Stephen Daedalus's wish to awake;[31] Bede and Matthew of Westminster recount the past in order to repeat it, for the past contains all that is most worthy, and nothing is worthy that is not in the past. The sins of the past can also provide instruction because they were sins in the past in the same way that they are sins in the present.

In this inventory of the icons of temporal mediation, three other items require at least some mention: relics, the Church, and the empire; indeed, the Church and the empire might be considered to belong to the set of relics. A cult of devotion to the bodily remains of apostles, saints, and martyrs, and then to objects associated with them, began soon after the passing of the generation of the apostles, as I have detailed in the previous chapter. By Augustine's time, the cult was well established and he attributes a number of healing miracles to the relics of martyrs. Gradually, the persistence of physical objects from the holy past into the present became the main object of faith, and the empowering foundation of churches. For Bede, the foundation of the English church takes place when the relics of "all the apostles and of divers martyrs"[32] are interred in the tomb of the English martyr, Saint Alban. The list of relics known to the Church proliferated as the sixth age drew into its second millennium, which it was never supposed to contain. The true cross, the lance, the crown of Christ—these became the main objects of devotion, replacing the Bible and even the Mass in the intensity and effectuality of the faith they evoked.[33]

The Church and the empire are generally understood in terms of the cultural coherence they conveyed to the medieval West (although the struggles between them were also the cause of the greatest chaos and strife), but they also mediated a temporal coherence. The Church was the great vessel of unity linking the past Christ with the present, both a relic itself and the living reliquary containing all relics. The medieval mind, as I have mentioned with regard to Henry VIII and Dante, saw the Church not primarily as a present institution but as a past mystery persisting into the present as an iceberg persists into the air. On its return from Avignon, the papacy moved from the Lateran to the Vatican in order to have its seat above the very bones of Peter. To the reliquarian imagination, this physical link was an important aspect of the apostolic authority. It may even be that among the many factors that eventually diminished the authority of the Church as a mystery, that attenuated the ability to revere the offices of the Church despite the rampant profligacy of the officers, the physical rather than the political removal to Avignon was most important: How could a bishop of Avignon be the patriarch of Rome, who derived his priority over the other bishops of Christendom by virtue of sitting in the apostolic see?[34]

The empire also was a mediation between the present and the Constantinian past. Roman cultural identification was an essential cement in the great Christian unity. The threefold thesis of each of the early national histories—Gregory of Tours's *History of the Franks*, Helmold's *Chronicle of the Slavs*, Saxo Grammaticus's *History of Denmark*, Jordanes' *Gothic History*, and others—is that this people has a distinctive and noble history; they have been brought into the flock of Christ; and they have been enfolded into the Roman state and Latin culture. Roman cultural identity had been an essential feature of Christianity since Eusebius. Dante even proposes that the Roman nation was the elect of God, chosen to prepare the world politically for the Christ, just as the Jewish nation had been chosen to prepare it theistically and morally. In his rhetoric, the Roman Empire becomes even more important a messianic vessel than the Jewish nation:

> Rationibus omnibus supra positis, experientia memorabilis attestatur; status videlicet illius mortalium, quem Dei Filius, in salutem hominis hominem adsumpturus, vel expectavit, vel quum voluit ipse disposuit. Nam si a lapsu primorum parentum, qui diverticulum fuit totius nostrae deviationis, dispositiones hominum et tempora recolamus; non inveniemus, nisi sub divo Augusto Monarcha, existente Monarchia perfecta, mundum undique fuisse quietum. Et quod tunc humanum genus fuerit felix in pacis universalis tranquilitate, hoc historiographi omnes, hoc poetae illustres, hoc etiam Scriba mansuetudinis Christi testari dignatus est, et denique Paulus, "plenitudinem temporis" statum illum felicissimum appellavit. . . . Admirabar

equidem aliquando, Romanum populum in Orbe terrarum sine ulla resistentia fuisse praefectum; quum tantum superficialiter intuens, illum nullo jure, sed armorum tantummodo violentia, obtinuisse arbitrabar. Sed postquam medullitus oculos mentis infixi, et per efficacissima signa divinam providentiam hoc effecisse cognovi, admiratione cedente, derisiva quaedam supervenit despectio, quum gentes noverim contra Romani populi praeeminentiam fremuisse, quum videam populos vana meditantes, ut ipse solebam, quum insuper doleam, Reges et Principes in hoc vitio concordantes, ut adversentur Domino suo, et unico suo Romano Principi.

(A phenomenon not to be forgotten attests the truth of all the arguments placed in order above, namely, that condition of mortals which the Son of God, when about to become man for the salvation of man, either awaited, or ordained at such time as He willed. For if from the fall of our first parents, at which point of departure began all our error, we survey the ordering of men and times, we shall find no perfect Monarchy, nor the world everywhere at peace, save under the divine monarch Augustus. That men were then blessed with the tranquility of universal peace all historians testify, and all illustrious poets; this the writer of the gentleness of Christ felt it meet to confirm, and last of all Paul, who called that most happy condition "the fullness of the time." ... I, in truth, at one time marvelled that without resistance the Roman people had become sovereign throughout the earth; for, looking merely superficially at the matter, I believed they had obtained sovereignty not by right, but by force of arms alone. However, after the eyes of my mind had pierced to the marrow thereof, and I had come to understand by most convincing tokens that Divine Providence had effected this thing, my wonder vanished, and in its place arises a certain derisive contempt when I hear the heathen raging against the preeminence of the Roman race; when I see people, as I was wont, imagining a vain thing; when, more than all, I find to my grief kings and princes concordant only in the error of taking counsel together against their Lord and His one Roman Prince.)[35]

Rome has here replaced Israel as the elect nation. Although Dante would not of course deny that Israel had been part of the *preparatio*, Rome has taken on the greater part. More important, while Israel has rejected its Messiah, Rome has accepted the revelation and has become both the center of the Christian Church and of the Christian nation. Therefore, not only the Church has received that election which God formerly visited upon the Jews, but Rome as city, as nation, as empire has also received it. And Roman cultural identity is a very flexible, inherently syncretic conception; it does not involve race, but rather conferred citizenship (in this it mirrors Christianity itself). Just as Christians are made sons of God by adoption, so can any people be included within the empire. The signs in one case are baptism, the sacraments, Scripture, and credal faith; in the other submission to the emperor, the Greco-Roman literary, philosophical, and legal tradition, and the Latin language. And as the Roman culture could cement the tribal cultures into a unity, so could it also cement the years. Orosius had presented the sack of Rome by Alaric as a minor episode in the fluctuations of the empire's fortunes. His theme was taken up (he and Augustine were the two chief *auctours*) by the medieval tradition. That there was a period without a Western emperor, or even an exarch, did not trouble the Middle Ages; for them the empire could have one emperor or two, be centered at Rome or Constantinople or be divided between them; all of these possibilities were allowed by the tradition, and either center of empire could be revived if it had fallen into abeyance. This is how Otto of Freising describes the crowning of Charlemagne as emperor:

> Anno ab incarnatione Domini DCCCI, ab Urbe vero condita MDLII, Karolus XXXIII regni sui anno a summo pontifice ablato patricii nomine, coronatus omni populo ter. acclamante: 'Karolo augusto, a Deo coronato, magno et pacifico Romanorum imperatori, vita et victoria', LXVIIII ab Augusto imperator et augustus vocatur.

> (In the eight hundred and first year from the incarnation of the Lord, and the one thousand, five hundred and fifty-second year from the founding of the City, Charles, in the thirty-third year of his reign, was relieved by the supreme pontiff of the title of patrician and was called Emperor and Augustus, the sixty-ninth in line from Augustus. As he was being crowned all the people shouted thrice, "To Charles Augustus, crowned by God, the great and peace-bringing emperor of the Romans, long life and victory!")[36]

Charlemagne's ascendancy is placed within both the Christian and Roman chronologies.[37] Otto notes that the coronation takes place in the thirty-third year of Charlemagne's reign, placing him in typical correspondence to Christ; elsewhere he has noted that Augustus and Christ are types of each other.[38] Charles is called Augustus and

Emperor of the Romans, and after this passage Otto frequently refers
to him simply as Augustus. Of later emperors the title Caesar was generally used.[39] There is no indication here that Charlemagne and later
emperors are not Roman emperors in exactly the same manner as
Augustus. The fall of Rome as a historical idea has its origins in the
Renaissance and is not fully developed before Gibbon. For medieval
citizens of both East and West, the Roman Empire had never fallen;
they were the Roman Empire.

I began this chapter by showing the forces of the medieval past in contention; I will end by showing them in harmony. Having separated the
various parts of the building from the matrix of the whole—and so
interdependent are they that they are difficult to separate, and may be
distorted in the process—I will now look at the matrix itself. If one
seeks a representative medieval history, the universal chronicle naturally reveals much more of its complete philosophy of time than the
particular chronicle of a nation, a war, an episcopal see. Of the Western medieval world-chronicles three stand out in their artistry and
comprehension, those of Joachim of Fiore, Vincent of Beauvais, and
Otto of Freising; it is the third that I think represents the city of the
medieval past in its completest form.

Otto of Freising's *Historia de duabus civitatibus*, compiled between
1143 and 1147, begins with a letter of dedication, written in 1157,
some years after the history itself had been completed, addressed to
the emperor Frederick I, Barbarossa:

> Petivit vestra imperialis maiestas a nostra parvitate, quatenus liber, qui ante
> aliquot annos de mutatione rerum a nobis ob nubilosa tempora conscriptus
> est, vestrae transmitteretur serenitati. Parui ergo libens et lubens vestro
> imperio tanto devotius, quanto regiae excellentiae convenientius esse con-
> sidero ob rei publicae non solum armis tutandae, sed et legibus et iudiciis
> informandae incrementum antiqua regum seu imperatorum gesta vos velle
> cognoscere.

> (Your Imperial Majesty requested of my humble self that the book which
> several years ago by reason of the beclouded condition of the times I wrote
> on the vicissitudes of history be now transmitted to your Serene Highness.
> I have therefore obeyed your command willingly and gladly, so much the
> more devotedly as I regard it as thoroughly in accord with your royal pre-
> eminence that you desire to know what was done in olden times by kings

and emperors, and to know this not only for the better protection of the
state by arms, but also for its better molding by laws and statutes.)[40]

"De mutatione rerum" does not mean mutation in the sense of
development or qualitative change; rather it designates the temporality and meaninglessness of human history. *Mutatione*, in Otto's doctrine, is the constant state of the universe, the aggregate of the petty
perturbations of human fortune and decay. Its sameness, as opposed
to difference, is attested by his intention that the emperor should learn
from it, as a gardener might learn from the seasons, not change, but
predictability. What the emperor should learn from this is the strength
and power of God, who changes monarchs and gives thrones to
whomever he will. For this reason, writes Otto, he will not merely give
events in their chronological sequence, but will weave them together
in the manner of a tragedy. There are, perhaps, two contradictory sentiments here: because of its unity, the past can provide precedents for
good governance, but it is also a tragedy that points beyond itself to a
kingdom not of this world. This is the historical sense of a man who
wrote of the Second Crusade—on which he went and in which he
played a leading and perilous part—that it did no good for the extension of boundaries, but was valuable only in that it saved men's
souls.[41]

The architectural scheme of Otto's history is laid out in a second
dedication to Frederick's chancellor, Rainulf:

> Quatuor principalia regna, quae inter cetera eminerent, ab exordio mundi
> fuisse in finemque eius secundum legem tocius succesive permansura fore
> ex visione quoque Danielis percipi potest.

> (That there were, from the beginning of the world four principal kingdoms
> which stood out above all the rest, and that they are to endure until the
> world's end, succeeding one another in accordance with the law of the uni-
> verse, can be gathered in various ways, in particular from the vision of
> Daniel.)[42]

Otto has two primary (and many intermediate) sources for the four
empires: Orosius, and Jerome's commentary on the Book of Daniel.
The four empires are, according to Otto, the Assyro-Babylonian, the
Medo-Persian, the Macedonian, and the Roman. This follows Jerome's identification rather than that of Orosius, who substitutes Car-

thage for Medo-Persia. The last empire, Rome, is to endure until the
end of time:

> Qualiter etiam regnum a regno subplantatum usque ad imperium Roma-
> norum fuerit, ostendi, hoc, quod de ipso dicitur, quia a lapide exciso de
> monte plenarie subvertendum sit, usque in finem temporum iuxta Meth-
> odium expectandum estimans.

> (I have shown how kingdom was supplanted by kingdom up to the time of
> the empire of the Romans, believing that the fulfillment of what is said of
> that empire—that it must be utterly destroyed by a stone cut out from a
> mountain—must be awaited until the end of the ages.)[43]

Rome is the *telos* of earthly history, the greatest of those secular
cities upon which God has chosen to bestow the scepter. And after
Rome is nothing within time, but only that apotheosis which is the
end and purpose of time. The scheme derived from Daniel places Otto
and his audience at the very end of time, on the last page of the confined, bound text of history, which is the only time and place that any
history that pretends to truth can declare itself to occupy. The end of
the ages (the plural has persisted into modern usage in a vague and
now meaningless sense) refers specifically to the last of Augustine's six
millennia, and this is the second scheme on which Otto's work is built.
The two schemes were never in conflict; Augustine's maps the history
of redemption; Orosius's or Jerome's the history of the earthly rule.
And so Otto stands in the sixth age, under the fourth empire, a citizen
of the Roman state, a bishop of the Roman Church. He dates the time
he sees from this magnificent vantage as so many years, before or after
the founding of the city, meaning Rome.

There is yet a third introduction to the *Historia*, another complexity
in the processional (an entry far more involved than the scaffolding to
the gate of Plotinus Plinlimmon's philosophy, and not at all intended
to be temporary)[44] into the history, a prologue to the first book. In it
is found yet another paradigm: the doctrine of the two cities. Otto
recounts how he has long pondered upon the things of time, and has
found nothing in them but vicissitudes ("vario ac inordinato"), and
holds that a wise man does not cleave to the things of time, and that
only by wisdom might one transcend and escape ("transeundum ac
migrandum") them. It is the distinction of a wise man not to be
whirled about like a revolving wheel ("Sapientis enim est officium non

more volubilis rotae rotari."), but to be firm. Accordingly, since all things are in constant and meaningless flux, and can neither be at rest nor afford rest, the wise man ought to depart from them to that city which remains at rest and abides for ever. This is the city of God, the heavenly Jerusalem,

> ... ad quam suspirant in peregrinatione positi filii Dei confusione temporalium tamquam Babylonica captivitate gravati. Cum enim duae sint civitates, una temporalis, alia eterna, una mundialis, alia caelestis, una diaboli, alia Christi, Babyloniam hanc, Hierusalem illam esse katholici prodidere scriptores.

> (... for which the children of God sigh while they are set in this land of sojourn, oppressed by the turmoil of the things of time as if they were oppressed by the Babylonian captivity. For, inasmuch as there are two cities—the one of time, the other of eternity; the one of the earth, earthly, the other of heaven, heavenly; the one of the devil, the other of Christ—ecclesiastical writers have declared that the former is Babylon, the latter Jerusalem.)[45]

This is the scheme of Augustine, but with slightly less subtlety; the cities stand in opposition; their interparticipation has disappeared. To these three paradigms of ages, empires and cities, Otto adds another, the transference of power from East to West, which he must have derived from Justin's synopsis of the no-longer-extant universal chronicle of Pompeius Trogus.[46] The transference of domination from East to West can easily be seen in the four empires (Trogus's thesis is a more detailed version of this), and Otto extends the movement within the Roman Empire. The power, he writes, was transferred from the city to the Greeks (Constantinople), then to the Franks, both Gallic and Germanic. Why the movement to Constantinople does not upset his paradigm, he does not state. This westward migration applies not only to earthly dominion, but also to philosophy. Learning began in Egypt (Plato is Otto's *auctour* for this), migrated to Greece, then Rome, finally to France and Spain. This migration is not a progression, but the transference always of the same knowledge, the same power, which gradually diminishes:

> Et notandum, quod omnis humana potentia seu scientia ab oriente cepit et in occidente terminatur, ut per hoc rerum volubilitas ac defectus ostendatur.

(And so it is to be observed that all human power or learning had its origin in the East, but is coming to an end in the West, that thereby the transitoriness and decay of all things human may be displayed.)[47]

In this prologue, Otto also gives a list of his *auctours*, which recapitulates the whole chain of influence that leads to medieval historiography: Pompeius Trogus, Justin, Cornelius (meaning Tacitus), Varro, Eusebius, Jerome, Orosius, Jordanes, but most of all, he adds, the great lights of the Church, Augustine and Orosius.

Here, in Otto's three-chambered entrance to his history, can be seen in confluence all the various contributory streams of influence that create the medieval sense of time. The whole forms a complex taxonomy of time, contains its perceived chaos within a salvific and supernatural providence expressed as a typological and numerological symmetry: two cities through four empires within six ages, as Otto writes of Christ's incarnation:

> Cum enim pro primi parentis, qui paradysi delicias postponens ex proprii arbitrii voluntate hanc terram maledictionis inhabitare maluit, culpa abolenda incarnari voluit, VI id etate potissimum fieri conveniens fuit, quia et ipsum VI die creavit.

> (For inasmuch as he willed to be made flesh in order to atone for the sin of our first parent, who, putting away the delights of paradise, preferred to inhabit the land of the curse at the caprice of his own will, it was most fitting that this be done in the sixth age rather than in any other, because he also created that first man on the sixth day.)[48]

God's providence is not only benevolent but also formal and aesthetic. From *The Iliad*, with its symmetries of numbered days and corresponding events, onward, Western literature has sought forms of balance and harmony that overcome the mere chronological process, the mere unwinding of the tale. In medieval history, the universe itself becomes that wrought object—the cup of Nestor, the shield of Achilles, the Grecian urn—in which all the elements are set in atemporal and simultaneous balance, in which every relationship is a "therefore," and not merely a "next." God, who symbolizes nothing, has filled the world with symbols about himself. The medieval symbolic process is precisely the opposite of that modern one that finds its most complete expression in Frazer's *The Golden Bough*. For Frazer the physical world is the archetype to which all ectypes point: Attis, Adonis, Osiris, Christ—all the avatars of the dying and resurrecting God—symbolize the vegetation myth, the seasons of the year's turning. To the medieval symbolist, God is the archetype and the physical world his ectype: God created the turning of the year, the seasons, to be a symbol of the death and resurrection of his son. Also, it would never occur to Otto that the shapes of time he discerns are a human invention; they exist for him not in the text of his book, but in the objective universe itself. This is one of the most important functions of the *auctour*: to provide ancient witness that these things are universally perceived, to free the present author from invention.

From the account I have given of Otto, and from the account he gives of himself, a reader might deduce several erroneous conclusions, false although supported by the evidence: that Otto exalts Rome as a type of the city of God, that he condemns it as the city of the world, that his process of decline constitutes a kind of entropy, that he is a historical pessimist. Otto's attitude to Rome is subtle: he does not identify it with his city of Babylon, a designation reserved for the pre-Roman power. Rome is to him a holy conduit even in its pagan form;[49] in its Christian era it is established as holy by the bringing into the city of the true cross by Helena, Constantine's mother.[50] No greater reliquarian consecration of the Holy Roman Empire is imaginable. Yet Otto takes the opposite part to Dante in the struggle between Church and empire. The decline of the empire mentioned in the dedication, which bodes the end of time, refers specifically to the Investiture Controversy, rather than to some general decline of power and culture. Otto admits, rather sheepishly, in his epistle dedicatory to the emperor Frederick, that now the empire is no longer usurping the power of the Church, he might take a rather brighter view of things. Otto believes that the empire should be completely subservient to the Church, and that it becomes, in a mystery, part of the Church, which has two roles, sacerdotal and royal.[51] It is the Church, in its most exalted conception, that inherits the earthly dominion.[52] The Roman Church and the Roman Empire become one inseparable, mystical body. Although Rome began in treachery, it is so sanctified by its Constantinian Christianization—predestined by God, who chose the city, while it was still in sin, to be his holy nation—that it ceases to be the earthly city, and becomes the city of God on earth. Thus, the city, in the sense of the Roman historians, becomes the city of God in the sense of Augustine (who would never have made such an identification), and the city becomes the Church:

> At deinceps, quia omnis non solum populos, sed et principes, exceptis paucis, katholici fuere, videor mihi non de duabus civitatibus, sed pene de una tantum, quam ecclesiam dico, hystoriam texuisse. Non enim, quamvis electi et reprobi in una sint domo, has civitates, ut supra, duas dixerim, sed proprie unam, sed permixtam tanquam grana cum paleis. Unde in sequentibus libellis non solum Romanorum augustis, sed et aliis nobilium regnorum regibus Christianis factis, cum *in omnem terram et in fines orbis terrae exi*erit *sonus* verbi Dei, tanquam sopita civitate mundi et ad ultimum plene exterminanda, de civitate Christi, sed quamdiu peregrinatur, utpote sagena missa in mare, bonos et malos continente, ceptam historiam prosequamur.

> (But from that time on [the Arian heresy], since not only all the people but also the emperors [except a few] were orthodox Catholics, I seem to myself to have composed a history not of two cities but virtually of one only, which I call the Church. For although the elect and the reprobate are in one household, yet I cannot call these cities two as I did above; I must call them properly but one—composite, however as the grain is mixed with the chaff. Wherefore in the books that follow let us pursue that course of history which we have begun. Since not only emperors of the Romans but also other kings [kings of renowned realms] became Christians, inasmuch as the sound of the word of God went out into all the earth and unto the ends of the world, the City of Earth was laid to rest and destined to be utterly exterminated in the end; hence our history is a history of the City of Christ, but that city, so long as it is in the land of sojourn, is "like unto a net, that was cast into the sea," containing the good and the bad.)[53]

This is why Otto's history is neither entropic nor pessimistic. As the worldly power and worldly wisdom tend always to fluctuation and decay, a new power enters into the world, God's salvific purpose, which, like the converging lines in an optical diagram, reaches its focal point at the Incarnation and ever after diverges outward as the Church. The tragedy of *mutatione*—and history is valuable as the source of exemplars of the principle—should teach contempt for the world:

> Verum quia regno decrescente ecclesia, ut dixi, bravium eternae patriae ac post vitae presentis laborem requiem adeptura, in presenti quoque in magnum montem crescens in magna auctoritate stare cepit, ea quae secuntur

... tam defectu rerum temporalium quam provectu spiritalium mundi contemptum prodentia in hoc opere dicenda restant.

(But because, as I have said, the Church, which is destined to obtain the glory of the eternal country and after the toil of this present life to attain rest, and at this very time is, besides, growing to mountainous proportions and has begun to reach great authority as the state declines, it remains for us in this work to relate the results that followed in consequence both of the failure of the temporal and of the advance of the spiritual, and which give rise to a feeling of contempt for the world. . . .)[54]

This contempt for the world does not mean, for Otto, a devaluation of the good and ordinary things of life: learning, friendship, sunlight, food. It is instead a contempt for the formless, plotless chaos of historical violence, of the lust for violence and domination for its own sake and seen as glory. By this measure, it is the later, chivalrous and courtly chroniclers, Froissart and Chastellain, who dispense pessimism (although they are not consciously pessimistic), for they endorse that world which Otto despises, a world in which perpetual warfare and slaughter and the lust for earthly glory become the highest of human achievements. For Otto the goal of history is its apotheosis, and he concludes his chronicle with the apocalypse, the dawning of the seventh age, which will never end. This apotheosis is, in a mystery, already accomplished by the enfolding of the empire into the Church, although the two-natured compound still walks in the land of sojourn. All evidence of decay, of strife, of the falling of those structures still left of the earthly city, becomes a testimony of the immanence of the eternal country.

Otto of Freising, Vincent of Beauvais, Joachim of Fiore, the great ecclesiastical historians of the high Middle Ages, achieve the fulfillment of that compensatory history begun by Eusebius. Time becomes, in them, the map or the book of time: the verbal structure transforms the experience of temporal absence into a plenitude of symbols; the darkened room has been hung with icons. Discourses, anatomies, taxonomies—the time between events has been made habitable by naming, and the time of sojourn, the sixth age of the world, made a perceivable unity by the profusion and elegance of its architects and *auctours*.

3

Antichrist and Antihistory

> Our lives could be a myth of captivity
> Which we might enter: an unpeopled region
> Of ever new-fallen snow, a palace blazing
> With perpetual silence as with torches.
>
> —GEOFFREY HILL, "Funeral Music"

The decline of the medieval model of the past and the rise of the modern is an untidy process, one that takes perhaps four hundred years to reach completion. There are, however, two intellectual changes, in themselves quite simple, that are the nodes from which a new historical configuration must emerge: the identification of the institution of the papacy with antichrist, a relatively sudden reversal of fifteen hundred years of religious and historical belief, and the rejection of the *translatio imperii*, the doctrine that the Holy Roman Empire was identical with its classical model. The reasons for these changes are immensely complex and overdetermined, as are the consequences that flow from them, the totality of which is the modern world, but the two changes themselves can be described with some hope of reasonable comprehension. These two form the isthmus connecting two great continents.

The origins of the identification of papacy and antichrist can be assumed to belong to the heterodox of the medieval period and possibly of late antiquity as well. The accusation is a logical one for any anti-Catholic group to make, and would fit well into the Gnostic schemes of inversion that constituted the speaking "wisdom among

66

the perfect" (1 Cor. 2:6). The identification may have developed contemporaneously with the papacy's own development as a distinct and definite institution, and (one might logically speculate) ought to have been held by the Cathars and various other heretical sects. But the question of the origin of the doctrine is largely irrelevant; what is important is the point at which the idea enters the *agora* of ideas, and into extant records. The Church was able to suppress the heterodox minorities and most of their writings; what matters is the dissemination of an idea, the instance the Church cannot suppress, the heterodoxy that threatens to become orthodoxy by the number and quality of its adherents.

Among the orthodox, Dante comes perilously close. In canto thirty-two of the *Purgatorio*, the triumphal car representing the history of the Church sprouts horned heads, and the beast, once formed, supports the great whore of Revelation 17:3.[1] But Dante is complex, elliptical, allegorical, and from his prose works it appears that he still maintained the distinction between the office and the office holder. Even so, the audacity of his allegory is startling.

The first writer to equate the pope with antichrist in plain words and to find an audience beyond a cultic sect was John Wyclif,[2] and from Wyclif on, the stream of transmission is no longer subterranean but obvious and seemingly inevitable, as is the historical revolution it necessitates. Some Wyclif scholars have suggested that he was, in this matter of the antichrist, interpreted rather too literally by his readers.[3] Certainly his initial identification must be seen in the light of his extreme philosophical realism. What he meant by the antichrist, goes the argument, was not the literal, prophesied, apocalyptic antichrist, but rather the spirit or form or universal of antichrist, which exists in all men to the extent that their wills oppose God. To Wyclif the universal was more real than the particular, and what a nominalist modernity would describe as an abstraction or a principle was more concrete to him than any empirical manifestation of it. Hus and later Luther (the latter an extreme nominalist) misread the type of antichrist into the antichrist of Revelation. If one were to look only at the progression of the identification through *De ecclesia*, *De blasphemia*, and *De apostasia*, this view would be licit, but *De potestate pape* renders it untenable. That a confusion between a realist and a nominalist antichrist, between the type and the person, takes place is, I think, undoubtedly true; but the process of hardening, perhaps of misinterpretation, occurs not between Wyclif and his followers, but within Wyclif himself.

In his early thought at least, Wyclif had no objections to the institution of the papacy as such, but only to its abuses and exaggerations. He distinguishes between three parts of the Church, one triumphing in heaven, another sleeping in purgatory, a third battling on earth.[4] The pope may legitimately be the head of the third of these, the Church militant, but cannot claim to be the head of the universal Church throughout time. The Church can have only one true head, who is Christ. Even the pope's headship of the Church militant must be constricted, as the Eastern churches do not acknowledge his headship.[5] In the light of this, the kinds of claims made by Boniface VII in *Unam sanctam*, to universal domination in all matters of salvation and of earthly rule, to an immediate control of heaven, purgatory, and hell, are an absurdity. Wyclif further believed that in order for the pope to exercise his legitimate authority, he must be an imitator of Christ, live and teach in accordance with the Bible, love God more than money, and refrain from simony, murder, and the like. The essence of Wyclif's propositions was that the office could not be separated from the man:

> Et stat quod aliquis solempnitate, ritu et reputacione humana sit reputatus Christi vicarius, cum hoc quod sit horrendus dyabolus, ut non est incredibile de Gregorio XI. et multis ei similibus. Nam si maritavit Raymundum nepotem suum cum herede Bolonie mediantibus decimis et bonis pauperum ecclesie Anglicane, si sustentavit iuxta fastum seculi multas parentum suorum familias, si redemit dignam captivitatem fratris sui et occidi multa millia hominum pocius propter secularem . . . nec finaliter fructuose penituerit, quis dubitat quin fuit perpetuus hereticus, nunquam caput vel membrum sancte matris ecclesie. . . . Hec autem non impono sibi, sicut faciunt superiores persone ecclesie, sed dico quod nec ipse nec alius est Christi vel Petri vicarius nisi relinquens mundanum ritum imitetur eos in moribus, et sic stat quod pretensus Romanus pontifex sit caput membrorum dyaboli.

> (And so one may be, by all human solemnity, rite and reputation, reputed the vicar of Christ, and yet be a terrible devil, as is not incredible in the case of Gregory XI and many of the same kind. For if a man used the tithes and goods of the poor of the English church to marry his nephew Raymund to a Bolognese heiress, and sustained the families of many of his kin in worldly pomp, and bought his brother out of deserved imprisonment, and

had many thousands of men killed for worldly gain . . . and did not finally
repent, who can doubt that he was a perpetual heretic, and never head nor
member of holy mother Church? . . . I do not impugn him [i.e., his
salvation], as do some ecclesiastical superiors, but I do say that neither he
nor any other is a vicar of Christ or Peter unless he relinquishes the ways
of the world and imitates them in death [to the world?], and thus such a
pretended bishop of Rome can be the head of the members of the devil.)[6]

Wyclif's condemnation of the papacy at this stage is not different
from those of Saint Catherine of Siena, of Dante, later of Savana-
rola—that some have been counted among the orthodox and others
condemned as heretics is largely a caprice of Fortuna. Wyclif, indeed,
asserted the need for a pope to lead the Church militant; he only
wished that pope to be poor and the spiritual model of Saint Peter,
and not to exceed his proper authority. In keeping with Wyclif's moderation of argument at this relatively early stage in his thought, the
references to antichrist in *De ecclesia* and *De apostasia* are very general, very realistic in the Scholastic sense. It is, writes Wyclif in *De
ecclesia*, the mark of antichrist to seek worldly domination, for he is
the ruler of the world, and if the clergy seek worldly domination and
wealth, they participate in the principle of antichrist, become his
members. This is repeated later, and associated with the abomination
of desolation standing in the sanctuary, from Daniel 9:27 and Matthew 24:15. The prophetic allusions are, I think, to be understood typically. The abomination in the sanctuary signifies the spirit of antichrist in the Church, in the form of worldliness, the lust for wealth
and power. But the passage does not point to any specifically historical
fulfillment, nor to an identification of the papal church and the antichrist. As a hyperrealist, Wyclif would designate as antichrist anyone
who desires to be the ruler of this world. He believes, of course, in a
proper civil authority, but this is to be exercised with moderation, as
a duty, and not by the Church. The claim to universal power, spiritual
and political, over every human being, is the principle of antichrist,
and it is this principle that Wyclif sees embodied in *Unam sanctam*.
To the extent that any man harbors such desires, that man is
antichrist.

In *De apostasia* the allusions to antichrist seem even more universal, and consequently less harmful than those in *De ecclesia*. He, or it,
is everywhere presented as the universal enemy of the Church. In *De
apostasia*, although antichrist is still a disembodied (or, more accu-

rately, variously embodied) universal, his characteristics move him precariously near to the hierarchy of the Church. He distorts true Christian teaching into its opposite, is a sophist, pretends to stand in Christ's place, and extorts money. But here again is no historical antichrist. The title is often used in the plural; for instance, all those who accept the unjust excommunications of the papacy are called antichrists. But the identification is moving inexorably nearer. *De apostasia*, written after the Oxford condemnation of Wyclif in 1381, is notable for its altered tone, its new bitterness in contrast to the moderation of the earlier treatises.[7] For the decisive equation, however, the incipient spirit of revolution, which was entering into the world, had to wait for the ninth book of Wyclif's *Summa theologiae*, *De potestate pape*. The treatise as a whole smokes with the same air of extremity that infuses *De apostasia*. In it a new theme, or even obsession, arises—that of the one antichrist, "unus Antichristus."[8] In this figure Wyclif merges the universal and particular antichrists as Dante does his thieves. The doctrine is introduced in the sixth chapter: a pontiff who claims to be the vicar of Christ while contradicting the way of Christ is the worst of antichrists. Antichrist may be defined in several ways: the whole body of the reprobate is antichrist, and each member of that body is also antichrist, but there is also a great antichrist ("magnus Antichristus"), who must arise from the clergy. The special sin of the antichrist (which antichrist is not really apparent here, but probably all three definitions are included) is hypocrisy. The antichrist has eleven characteristics: he claims to be a follower of Christ, yet is the opposite of Christ, and leads others astray; he does not follow Christ's poverty, but loves wealth and power, and claims dominion over the whole world; whereas Christ has given the one true law expressed in Scripture, the antichrist proliferates new laws contrary to Scripture; Christ chose poor men to be his disciples, but the false one ("Pseudo") chooses only the illustrious; Christ sent his disciples to preach throughout the world, but the false one dwells in a rich palace and cares nothing for preaching; Christ taught his disciples to submit to tribulations and injustices, but the false one teaches his disciples to fight for earthly dignity and dominion; Christ forbade his disciples to take the sword, but the false one hires soldiers with the offerings of the poor; Christ would not call down fire upon the Samaritans, but the false one curses all those who will not submit to him; Christ refused to judge and submitted to judgment, but the false one pretends to judge the whole world; Christ hid his deity in incarnation, but the false one claims to be God and thus fulfills Saint Paul's prophecy of the man of sin; finally, Christ allowed his disciples no gain from their ministries, but the false one sells all prelacies for money, to those who only wish to make money by the sale of supposed salvation.[9] The theme is again taken up in reprise in the twelfth chapter, and here the identification becomes explicit. Of all those who are antichrists, there is one who is greater than all others, he who is prophesied in Revelation and in Daniel:

> Michi autem videtur quod nichil plus concordat cum litera vel ebdomali-bus Danielis quam exponendo hoc de pseudopapa, qui est potissimus Antichristus. In hoc enim est abhominacio desolacionis, et supra exposui.

> (Although there is little agreement concerning the seven days of Daniel, one may see how well the pseudopope, who is chief Antichrist, is pictured here, how greatly, indeed, the abomination of desolation is fulfilled in him.)[10]

The pseudopope stands in the holy place, the Church, and claims the worship due to Christ, commanding all to call him holy father ("pater sanctissimus"). And by his perversion of Christ's doctrine, he lays the Church to waste. Thus is fulfilled the phrase, "the abomination of desolation stands in the holy place [abhominacio desolacionis stans in loco sancto]."[11] Wyclif does aver that not every pope is antichrist—indeed many popes have been saintly—but the endowed, the rich pope is antichrist (here is a note of the universal again) and the antichrist will be a pope.[12]

What seems to be an inevitable convergence in the development of Wyclif's thought is completed in these passages of *De potestate pape*. Wyclif began by criticizing the corruption of the papacy, not the institution itself, and comparing that corruption to a universal antichrist, a Platonic (or, rather, Neoplatonic) form in which all opposers of God participate, and even all followers of God participate to the extent that they oppose God. By a gradual transference of terms, a lateral sliding of the real into the nominal, of the universal into the particular, all of this aided by the growing extremity of his anger, he comes to condemn the papacy itself, and to equate it with the single, prophesied, end-time antichrist, the vicar of Satan on earth. In so doing, he reverses the central tenet of thirteen hundred years of Christian history—quite simply he makes the cementing belief of the entire Western world from Eusebius to Wyclif's own present into its opposite, into a mirror world. The whole structure of history as developed by Eusebius, Augustine, and Orosius centered on the Church as the true repository, the reliquary, of the apostolic tradition. It was the chief of those icons of compensation that bridged the abyss between the believer and the Christ he had never known, both the abyss of time, through the doctrines of apostolic succession and of relics, and the abyss between earth and heaven, through the doctrine of the real presence. The idea of a history both linear and static, in which no qualitative change could occur, no new true thing be discovered, was a concomitant matrix, a time that was no time in which the Church lived, and in itself a compensatory icon, providing a temporal home, a medium almost tangible, in which each generation could feel intellectual communion with the generations before it, even to that sacrifice that was the ultimate icon, tangible sign of intangible covenant, the world's circumcision.

But the whole system was fragile. There is a kind of epistemological gravity that makes all systems, structures of balance, all attempts to describe the world as being just so (and not thus), and all attempts at completion tend to collapse. Any intellectual structure that pretends to completion can be destroyed by new information. The delicate medieval interplay of Christianity and history claimed to such completeness, holding as one of its essential doctrines the impossibility of the new. If Wyclif were right, that is, if he were to win adherents, among those adherents, the past would have a new shape. What he had discovered, to any who were persuaded by him, was a new truth, a secret hidden by God in time, reserved for the last age, a secret that reversed the meaning of the past and revealed all things to be their opposites.

Wyclif did not open the book of his ideas' historical implications, but he does reveal some extraordinary, almost accidental, information about his view of history, and by extension, perhaps an unraveling thread in the intellectual fabric of the fourteenth century. In *De apostasia*, he states that errors now abound in the Church because it is in the second millennium after Christ:

> ... perturbat ecclesiam. Ideo ulterius notandum quod in secundo millenario matris nostre, quo solutus est sathanas ut dicitur Apok. XX.

> (The Church is troubled because our mother is in her second millennium, for which reason Satan is loosed, as is written in Apocalypse 20.)[13]

In two other passages Wyclif introduces the idea of the second millennium: the next occurs in a condemnation of Innocent III's doctrine of the Eucharist, a doctrine that Wyclif believes to be an apostate neologism:

> ... nec est credibile quod ista quidditas hostie latuisset autores scripture et cunctos sanctos episcopos et doctores per mille annos et amplius; et secundo millenario quando solutus est pater mendacii, et clerus excidit a religione Christi, sit ista veritas fidei revelata.

> (It is not credible that what was utterly unknown to the authors of Scripture and all saintly bishops and doctors for a thousand years and more, should be revealed to the faithful in this, the second millennium, in which the father of lies is loosed, and seeks to destroy the Christian religion.)[14]

This is precisely the form of argumentation employed by Saint Bernard of Clairvaux and by Henry VIII, cited at the beginning of the previous chapter. All true doctrine is original doctrine, and, within the terms of argument that are the necessary correlative of the medieval idea of time, to prove the novelty of any idea is to disprove the validity of it. What is different here is the idea of the second millennium. Wyclif goes on to say that Mahomet and Sergius did less harm in the first millennium, than have the heresies of the papacy in the second. The third and final passage in *De apostasia* in which the second millennium is mentioned occurs in the context of a discussion of Augustinian history, to which Wyclif is intellectually devoted.[15] Writing of the thousand years of the twentieth chapter of Revelation, in which Satan is to be bound, Wyclif cites Augustine's opinion that the thousand years represent the age of the Church, the sixth millennium, in which Christ was born, which was fitting, again according to Augustine, because Adam was created and fell on the sixth day. This sixth millennium now being over, Satan is loosed and heresy stands in the heart of the Church.

There are two points to be noted here: first, that Wyclif is thoroughly in accord with the spirit of medieval history, both in his condemnation of the new and in his acceptance of Augustine's six millennia; second, that something is radically wrong. If the age of the Church is the sixth millennium, the last age of earthly history, and the seventh is to be the eternal sabbath without end, the future heavenly state after the Second Advent, then what is this second millennium after Christ, in which Satan is no longer bound? Augustine gave completion to Christian history, both symbolically, the seven ages of the world corresponding to the seven days of creation, and epistemologically, from the viewer's standpoint at the comparative end of time, in which all things were complete, except only Christ's return.

In Wyclif, Augustine's history works against itself, and against its provisions of comfort for the believer. The problem is, quite simply, time. Christian history was necessitated in its origins by Christ's failure to return within one, within two, within three generations. Augustine filled the abyss of extension with symbolic architecture; his sixth age postponed the return for a thousand years, alleviating the agony of expectation, making Christ past and Christ future infinitely far and near. The system contains, however, a single flaw that Augustine could not have anticipated: what happens when the sixth millennium is over, and Christ still does not return? True, the millennium does not have to be strictly a thousand years in Augustine's thinking, but this is what the word suggests. Wyclif is more than three hundred years into the seventh millennium after the creation, the second after Christ, and the seventh day has not yet dawned. Medieval history transformed an infinite abandonment into an indefinite expectation. Augustine's ideal millennium could only be stretched so far. This second thousand years was something unforeseen, a gigantic parenthesis opening up in the universal and universally accepted scheme of time, and the system, like all systems, had a very limited tolerance for the unforeseen. The creeping, inexorable extension of time after the year 1000 gradually strained the fabric of history. The sacred geometry of time lost its completeness and proportion. The age of the Church should have been followed by the actual and intimate presence of God, the face-to-face after the dark glass. Instead it was followed by more extension, by more of the dreary, plotless, bloody history of the city of men, further away from God-among-us, an even darker glass. Wyclif searched the Apocalypse and discovered an age of Satan.

As I have stated, changes in the historical model are overdetermined, and neither Wyclif nor his reception can be separated from the subservience of the Avignon papacy to the French monarchy, from the consequent and very worldly patronage of John of Gaunt, from the Great Schism during which every Christian people was both blessed and cursed by the rival popes (Wyclif welcomed the schism as a sign of the preposterousness of the papacy's—or papacies'—claims), nor from the almost industrial system of scribal transcription that obtained in the fourteenth century, a prelude to the printing press. Rhetoric cannot be separated from the physical and political conditions in response to which it is developed. Yet all of these things do not exist except as traces in the texts that describe them. For the present, the text is the primary evidence, and physical, political, and economic conditions are only the secondary conceptions, abstracts that may be derived from the text.

Leaving aside, therefore, the econo-socio-political causes, the trail of rhetoric is quite clear, and is indicated by the fact that all four of the Wyclif works that I have used above are extant only in Bohemian manuscripts.[16] The trail of influence leads directly from Wyclif to Jan Hus and his mentor-disciple Jerome of Prague. Hus's *De ecclesia* is nothing more than a précis of Wyclif's *De ecclesia*, with references to peculiarly English subjects removed, and passages from *De potestate pape* interpolated, so that Hus's work is a kind of abbreviated portmanteau of the two Wyclif tractates.[17] Intellectually, Hus is entirely Wyclif's creature and adds nothing of his own. As a disseminator, however, he was immensely important, spreading Wyclif's ideas throughout the empire while they were being posthumously suppressed in England. And Hus received a sympathetic although qualified reading from Martin Luther.

Although Luther wrote no systematic historical work (save the tiny and essentially private *Supputatio annorum mundi*) he is undoubtedly (with Petrarch) one the two most important formulators of historical ideology between Augustine and the eighteenth century. In asides and fragments Luther framed an entirely new and revolutionary way of viewing the past; in him the new system is essentially complete; to the Protestant historians—Sleiden, Illyricus, Foxe—remained only the task of formulating the Lutheran system as systematic and chronological history.

The *Supputatio*[18] is a universal chronological table, with parallel columns setting out biblical and secular events, dates and millennia, and general annotations. The whole begins with a note that human history will last for six millennia, corresponding to the six days of creation, the seventh millennium being the eternal sabbath of the saints with God.[19] Luther further divides these into two millennia before the law, two under the law, and two under the Messiah. Here is the persistent force of Augustine's historical imagination, reaching through a thousand years of the history he had created. No one, in this whole period, was capable even of imagining a non-Augustinian history, and Luther is not an exception in this. The *Supputatio* is Luther's addition to Augustine's six ages and to the Africanus/Eusebius/Jerome chronicle tradition, and it demonstrates the absolute intellectual hegemony that these two systems together exercised. But Luther, although he could not think outside his intellectual heritage (no thinker can), was able to modify it in unforeseen ways (unless foreseen by Wyclif). Luther's millennia do not exactly coincide with those of Augustine. Whereas Augustine's millennia are symbolic wholenesses divided by changes in divine dispensation, Luther's are literal epochs of a thousand years, divided both by divine event and by date. In Luther's calculation, the first millennium extends from the expulsion from Eden to Noah, the second from Noah to Abraham, the third from Abraham to the post-Solomonic division of the Jewish kingdom into the two states of Israel and Judah (a strange site on which to place a milestone in salvific history), the fourth from the division of the kingdom to the resurrection of Christ, the fifth from Christ to around A.D. 1033, with no particular demarcation point other than date, the sixth from this time until the end. The fifth millennium is the age of Christ; the sixth is denoted variously as the age of the pope, of antichrist, or of Satan.[20] Such references abound in the annotations to the events of this last millennium: for instance, the Council of Constance is entered simply as "Concilium Satanae," the council of Satan.[21]

As Luther's only "history," the *Supputatio* is an obvious point from which to begin an investigation of Luther's historical thought, but it is also, in many ways, the wrong point. Drawn up by Luther only five years before his death, it is the final link in a chain of logical propositions and consequences. Luther began his career as a reformer with a set of propositions about doctrine; these led him into opposition toward the papacy, and that opposition—doctrinal, then rhetorical, and potentially physical, had the Church been able to lay hands upon him—led him to a new understanding of the position of the papacy within the Church, and this in turn necessitated a radically new view of tradition, and of the nature of the Church within time. The purely theological elements of Luther are beyond the scope of this present work, but his doctrines of the papacy, of tradition, and of the Church are the three determinants that require, as a matter of necessity, a new history, both from Luther himself in the *Supputatio*, and from the Protestant historians who so massively expanded on its sketch. It is with the papacy, then, that an understanding of the new history must begin.

Compared to Luther, both Wyclif and Hus were paragons of moderation. Both accepted the idea of the papacy in some form—providing that the throne's occupant be holy and exercise a moderate and conciliar leadership rather than an autocratic dominion—but condemned its practice and the exaggeration of its authority. Luther, however, came to see the entire papacy as a Satanic invention from its very beginnings:

Denn wir lassen das Bapstum nicht sein die heilige kirche, noch ettwa ein stucke dauon, ond konnens auch nicht thün, Sondern es ist der wuste grewel ond Endechrist, der feind ond wider wertiger, der die kirche, Gotts wort ond ordnung zerstoret, ond sich selbs wider ond druber sekt als ein Gott ober alle Gotter, wie Daniel ond .S. Paulus geweissagt haben, Ond die weil es nicht sein kan, das wir oder die heilige kirche, sich leiblich scheide oder absondere, oon dem grewel Bapstum oder Endechrist, bis an den Yüngsten tag (Denn der grewel sol ond mus (wie Christus leret) nicht ausser, sondern enn dem tempel Gottes siken, ond das Bapstüm nicht ausser, sondern enn der kirchen sein) so mussen wir doch ons wissen geistlich ond mit rechtem oerstand, oon ehm zu scheiden, ond ons fur seinem oerstoren huten ond bewaren, damit wir im rechten glauben Christi rein bleiben, ond wider sein geschmeis ond onzifer ons wehren ond oerteedigen.

(For we claim the Papacy not to be the holy Church nor any part of it, and we are unable to cooperate with it. Rather it is the horrid abomination and the Antichrist of the end, the enemy and adversary which devastates the Church, God's Word, and order, and sets itself over and against them like a god over all gods as Daniel and St. Paul have prophesied. And while we or the holy Church are not able physically to separate or detach ourselves from the abomination, Papacy, or final Antichrist until the Day of Judgement—for as Christ teaches the abomination should and must stand not outside but in the place of the Holy, and the Antichrist must sit not outside but in the Temple of God, and the Papacy must not be outside but in the Church—so we must continue to recognize ourselves spiritually and with a right understanding separating ourselves from him, guarding and preserving ourselves from his devastation. Thereby we remain pure in the right faith of Christ, resisting and defending ourselves against his excrements and vermin.)[22]

If the current papacy is neither a God-ordained leadership, nor even a human institution convenient for good governance and order, but a

Satanic plot imposed upon the Church to lead it astray, a sign not of God's love for, but of his anger against, the Church,[23] then Luther must answer the question of when this Satanic papacy began. There were, he believed, several stages in the corruption. Constantine emerges as one of the villains because he brought the Church into imperial politics; another is Gregory the Great, about whom Luther was ambivalent. He viewed Gregory as the last bishop of Rome, and commended him for rejecting the title of *universalis episcopus* and for a saintly life; in debit, however, Luther saw in Gregory the great extender of Roman influence and regarded his teachings as idiocy (Saint Peter, said Luther, was the last pope to teach anything true). The turn to corruption consistent and Satanic, however, dates to the granting of the title to Boniface III by the murdering emperor Phocas. Yet in a sense this progression does not matter; it is part of the larger fall into corruption that begins immediately after the removal of the apostles:

> Sic in Ecclesia novi testamenti statim sequuti sunt Apostolos Haeretici. Item Episcopi, qui non norant Dominum, postea Monachi, et tandem totus Papatus, et universa abominatio stans in loco sancto. Atque omnes hi clamaverunt: Serviamus Deo aliquanto ardentius et maiore pietate. Multa enim reliquerunt Apostoli addenda Ecclesiae, patres parum devotionis et ceremoniarum habuerunt, accumulemus non plures ritus et cultus. Ad hunc modum auctae sunt ceremoniae in Ecclesia, miscente Diabolo vera falsis, et ruentibus posteris paulatim in deterius.

> (Thus in the church of the New Testament heretics immediately followed the apostles. Likewise bishops who did not know the Lord. Later came the monks, and finally the entire papacy and the whole sacrilege stand in the holy place. And all these cried out: "Let us serve God somewhat more ardently and with greater piety! For the apostles overlooked much that must be added to the church; the fathers had too little in the way of devotion and ceremonies. Let us accumulate more rites and acts of worship!" In this manner ceremonies were increased in the church, for the devil mixed truth with falsehood, and the descendants gradually rushed into what was worse.)[24]

Here, then, is the second of Luther's inversions of his inherited historical model. Christian historiography developed a traditionalist, or what Collingwood has called a substantialist, view of time.[25] Substantialism holds that only the eternal substances are true and worthy of knowledge, whereas the accidents, the particular and shifting events of the temporal world, are mere impediments to the knowledge of the substances. Collingwood locates substantialism in Greek metaphysics, and regards it as being ultimately incompatible with historical thought. This latter point depends of course on what one means by historical thought, and Collingwood is here universalizing, that is, substantializing the peculiarly modern assumption that history is about the dynamic, the fleeting, the changing, the particular, the unsubstantial. Thus what modern scholarship calls "historicism" has become a method of deconstructing the truth content of any text by demonstrating its place within a particular temporal system, which is past and therefore invalid.

Medieval historical thought made substantialism and traditionalism identical. The tradition can be defined as everything that a present generation receives from the past. This constitutes the valid and true, the substantial body of knowledge. Everything, therefore, that it is possible to know is already known, and all knowledge is available by the method of consulting the tradition. Tradition accumulates by denying its own accumulation, by declaring its innovations to be ancient and original. The result is a cognitive timelessness, without which Christianity, in its medieval form, could not exist. Luther's attack is precisely on this medieval concept of tradition. Yet the attack is made in medieval terms: what is wrong with tradition is that it is not original, but a novelty, unsubstantial, the trick of the devil.

Eusebius and Augustine rescued the Church from time, from the abyss between Christ and now, from the possibility that the Church's supposed knowledge of its past Savior was nothing but a regression of degraded simulacra, by accumulated error impossibly removed from the original; Luther declares that the unstated, suppressed fear that haunts early Church historiography is the truth, and declares it with a kind of glee. Whereas the early Church's epistemology committed it to asserting the continuity of present belief with past belief, Luther's rejection of current Church doctrine commits him to an assertion of its discontinuity with original truth, and this requires a violently dynamic history, a precipitous fall into unknowing. What all have held true for fifteen hundred years Luther declares a degraded simulacrum, a parody of Christian doctrine that, with the long accumulation of error, has become almost the opposite of the original:

> Non transferri terminos a prioribus positos est nihil addere doctrinae ab Apostolis traditae velut melius consulendo rebus conscientiarum. Et hunc

locum fortiter pro se iactant Sophistae et Pontifices, dum pro suis statutis
et consuetudinibus clamant non esse transferandos terminos, quos consti-
tuerunt patres. Per patres intelligunt autem suos Pontifices et Doctores non
autem Apostolos. Sic allegorica sententia suas fabulas stabiliunt interim
non videntes, quam ipsi sint primi et soli omnium (etiam allegoria stante)
transferentes terminos non modo fidei et spiritus per priores, Apostolos et
Christum in Euanagelio, positos sed etiam suos ipsorum a suis prioribus et
a seipsis positos, cum id sit umum eorum studium, ut leges legibus mutent,
accumulent et, ut ille ait, leges figunt precioque refigunt, ut videantur leg-
islationem vice ludi ac tesserarum habere, quibus ludant in conscientiis
hominum, et tamen aliis obiiciunt, ne transferant terminos, quos priores
posuerunt.

(The landmarks are not to be moved from where they have been placed by
the former dwellers means that we are to add nothing to the doctrine transmitted by the apostles, as though one could give better advice in matters of
conscience. This passage is also boastfully claimed by the sophists and pontiffs, who clamor concerning their statutes and customs: that we are not to
shift the boundaries that the fathers set up. By "fathers," however, they
understand their own pontiffs and doctors, not the apostles. Thus they support their fables by allegorical saying. Meantime they do not see that even
if the allegory were valid, they themselves are the chief and even the sole
movers of landmarks, not only the landmarks of faith and the Spirit, set up
through the former owners, the apostles and Christ in the Gospel, but even
their own landmarks, set up by their predecessors and by themselves. Their
one endeavor is to change laws by laws, to heap them up, and, as the saying
goes, to make laws and for a consideration to remake them. Thus they seem
to use lawmaking in place of a game and dice, to play with the consciences
of men; and yet they reproach others for changing the landmarks which the
fathers have set up.)[26]

The distinction, the boundary, that Luther draws here is precipitous. Only the apostles were capable of the truth; only they had the
authority to establish doctrine; all after them have been in error. The
early Church saw no sharp distinction between the apostles and the
episcopate the apostles founded; Luther makes an absolute distinction. For the living continuation of a tradition that saw the New Testament as one part of the Christian endeavor to speak about God, he
substitutes the single text alone, separated from the corrupt and fallible
Church. The tradition of the Church allowed for speculative theology;
Luther will allow only the hermeneutic exposition of the text, and
even this cannot be speculative as the Bible interprets itself: its mean-

ing is obvious and needs no tradition of interpretation. God is encountered, therefore, not in the body of believers, nor in history, nor the Mass, nor in relics, but only through the preached word, the disembodied text.[27] Where Eusebius and Augustine had made a home for the Church in the world, had alleviated the agonies of abandonment and expectation, Luther exalted these agonies as the only matrix in which faith can thrive. He asserted that the Church could only be true under martyrdom and tribulation,[28] that its visible form of papacy and episcopate and priesthood was a wicked distortion visited upon it because of God's anger,[29] that the past was irrecoverable and the *imitatio Christi* impossible,[30] that the present Church existed in an age of Satan unbound, and that the Church itself was Satan's instrument in the world.[31]

Luther's new history is still substantialist (as all Western history has been until relatively recent times), but it disagrees with medieval history as to what constitutes the substances. Eusebian and Augustinian history sought to redeem the accidents, to validate the Church's worldly as well as its mystical existence, to sweep up into God's salvific purpose the city of men as well as the city of God, to redeem even the pagan past as *preparatio*, to make Rome "that Rome where Christ is a Roman."[32] By contrast, the Lutheran vision of time, the Church, and the world is marked by its exclusion. It is the opposite of medieval syncretism: instead of books, the single book; instead of the past, tradition, as the source of truth, a great division driven between the primitive apostolic Church, lasting only one generation, and the present; instead of a universal, catholic, apostolic Church, a hidden and persecuted and invisible body of the few faithful within the apostate. Eusebian history, while it exalted martyrdom, had sought redemption even for the persecutors, and had rejoiced in the ending of persecution with official Rome's conversion; Luther exalted martyrdom as the only true condition of the Church, rejoiced in its onset—and this was to be a different form of martyrdom, one imposed not by the pagans but by the church apostate on the church faithful. And Luther's idea of the hidden Church was closely related to the idea of a hidden God, a *larva Dei* concealed as though by a mask,[33] who reveals the wise things always in the forms of foolishness, the truth hidden under the lie.[34] If the Reformation was a movement of great religious revival, it also contained within itself the opposite tendency: toward Deism and the withdrawal of God from human experience.

Just as Luther was no less concerned with substantialism than was medieval historiography, so he was no less afraid of neologism; indeed the Lutheran system is much more fragile in this respect. Luther, for instance, parallels Eusebius's concern that even the Incarnation might be suspect because it is a new thing: he asserts that the New Testament precedes the Old and is therefore superior to it, and that the Old Testament is called old not because it was first but because it has ended.[35] This assumption that the older, not the newer, is the greater unites Luther with the ancient, not the modern, world; similarly, his assertion of a dynamic rather than static history since Christ is not an approval of dynamic history but a condemnation of it. Where medieval history condemned innovation and denied that it had occurred, Luther condemned innovation and saw it everywhere. What the past had called tradition Luther called innovation. His own project was precisely the unity with the past that the Church had always sought, but he condemned all of the icons save one. Luther saw difference instead of sameness in time, and a yawning gap between the present and the primitive Church, whose distinct identity he found in Lorenzo Valla.[36] And at the heart of the primitive Church was the Bible, the single, original icon given directly by God, *sola scriptura*. This is why the Lutheran system is ultimately more fragile than the Eusebian/Augustinian: it relies for mediation upon only one icon, the book, now made identical with the Word of God. Christ at last had become fully identical to the words about Christ. This icon must be declared immutable, entirely without innovation, for if this falls then God becomes finally and utterly unknowable. But the history of unity, which was one of the most important of icons, is replaced by that dynamic history of difference, which is no icon, but a void, in which the decontextualized Word might, perhaps, shine more brightly. In medieval terms Luther's history is an antihistory, a great parenthesis driven between the present and the primitive Church.

The medieval mind could experience absolute intellectual identification with all past generations: the body of truth had been known to all; the Protestant mind could experience none: between the apostles and the Reformers almost all had been in error. The identity of the true Church was a secret hidden in time by the hidden God, revealed only to the last generation. For the first time since Eusebius a new thing was revealed (though it must be declared ancient and original, as all substantial things are), a newly discovered truth had superseded past error, and with the idea of supersession, time, in the human mind that experiences time, begins to move. So the restoration of true Christianity is a sign that the Second Coming is imminent: the last millennium plummets to conclusion:

> Nisi quod credo nos esse tubam illam nouissimam, qua praeparatur et praecurritur adventui Christi. Ideoque utcuncque sumus infirmi et parum sonamus coram mundo, Tamen magnum sonamus in conventu Angelorum coelestium, qui sequentur nos et tubam nostram, Et sic finem facient, Amen.

> (I believe that we are the last trumpet which prepares for and precedes the advent of Christ. Therefore, although indeed we are infirm and scarcely do we sound in the presence of the world, yet we sound greatly in the assembly of the heavenly angels who will follow us and our horn and thus make the end.)[37]

Wyclif and Luther made new history in scattered proclamations; both were interested not in history itself, as a systematic science or myth, but in history as an adjunct to theological polemic. Their historical pronouncements were isolated stilettos, asides and footnotes on the periphery of the antipapal polemic, points whose tiny area concealed and concentrated formidable mass. The radical theology demanded a radical history; Wyclif and Luther understood this, and became historians where need dictated. Luther combed the histories looking for evidence of the falsity of papal claims. And the new history is complete in Luther in implication (although Calvin adds greatly to doctrine, church governance, and liturgy, he contributes no necessity for new history beyond Luther's); it awaited historians to draw out those implications into full coherence.

The *Magdeburg Centuries* (so called because each volume covered a century), compiled between 1559 and 1574 by a committee of Lutheran scholars under the leadership of Matthias Flacius Illyricus (with notable contributions from the English Marian exile John Bale), was the great continental effort to write a Lutheran past. John Foxe's first attempts at his *Acts and Monuments* also belong to the period of the Marian exile. The other great expositions of Protestant history belong to England of the Elizabethan settlement: Matthew Parker's *De antiquitate Britannicae ecclesiae* and the later editions of Foxe. Of these, Foxe's is the greatest work and benefited from extraordinary propagation (it was commanded by royal and ecclesiastical decree to be displayed in the hall of every episcopal palace in England, and was almost as commonly displayed for public reading in churches as was the Bible itself) and, in consequence of this and of its peculiar attractiveness to the ideological moment, exerted an extraordinary influence. More than any other book, it is to the *Acts and Monuments* that one must turn in order to see what the past looked like to the post-Lutheran world. By one of those delightful accidents of biography, John Foxe seemed destined from birth to be of the Reformation: he was born in 1517, the year that Martin Luther nailed his theses to the Wittenberg door, and in Boston in Lincolnshire, the symbolic geographical center of the Puritan migrations to the New World.[38]

Foxe's history, in its ultimate form,[39] begins with a delineation of difference almost shocking in its intensity. Foxe cannot conceive of his history existing in its own right; its entire identity and structure is derived from its character as a *history against*, echoing the titles of the apologetic histories of the early Church, though not against the Gnostics or the pagans, but against the church of Rome. After a phalanx of letters dedicatory and prefaces—including one "To All the Professed Friends and Followers of the Pope's Proceedings,"[40] a condemnation, containing an exposition of the thirteenth chapter of Revelation proving the pope to be the second beast, and "Four Considerations Given out to Christian Protestants, Professors of the Gospel; with a Brief Exhortation Inducing to Reformation of Life,"[41] a flattering suasion to further good faith—comes the first heading of the history proper: "And first, the difference between the Church of Rome that now is, and the ancient Church of Rome that then was."[42] To anyone steeped in the medieval conception of time, the disjunction is shocking. Foxe posits a past and a present that have absolutely nothing to do with one another. "And first"—before any description can be given of this history, the vital difference must be asserted. "Difference," the third substantive word in the history, preceded only by the abrupt urge to immediacy of the first two, is Foxe's own, I do not impose upon him the term of any deconstructionist theory; indeed, the opposite is true: Foxe has imposed the word and the concept on all who come after him. The repetition of "Church of Rome" emphasizes in which dimension the difference is to be located: it is not a difference in the nature of "Church" nor of "Rome"; that is, no two churches are to be distinguished, for Foxe will admit of only one Church, nor is a spatial distinction to be made between two sees: the church of Rome as opposed to the church of Constantinople, for instance. The difference is temporal only, made emphatic by the contrasting, slightly altered echo: "that now is . . . that then was." And with these two phrases, the medieval temporal unity is destroyed, reversed, spun about a medial plane and made its own opposite and inverted reflection.

The process of rhetorical transformation, as I have traced it, has taken two centuries from Wyclif to Foxe, but the wide dissemination of the belief in the antichristian nature of the papal church—and with it the consequent necessity of a disjunctive history—dates only to Luther. The speed of change can be appreciated if one considers that Foxe reached full adulthood as a Catholic, apparently of complete orthodoxy, at least externally, for such would have been required for his election as a fellow of Magdalen, Oxford, in 1543. Although Henry VIII had been declared head of the Church in England in 1533, all forms of Protestantism were vigorously persecuted until the accession of Edward VII in 1547. Foxe was expelled from his fellowship on July 22, 1545, for the cause of heresy, that is, Protestantism.[43] Foxe's convictions in this period are unknowable, but a very marked change in his external conversation and profession must have taken place between 1543 and 1545. His conversion to Protestantism may not have been complete at this time, and he may not have realized its historical implications, but he must have done so by the 1550s, when he began work on the first version of what would become the *Acts and Monuments*.[44] The reversal of historical vision, from union to difference, must have taken place in a decade at most for Foxe, and his experience must have been replicated by almost all who followed the movement for reform; the past changed its shape, revealed the beliefs held universally by the Western world for fifteen hundred years to have been falsehood, in a single generation.

The times of the Church, writes Foxe, have been "horrible and perilous from the beginning, almost," but especially, "since Satan broke loose."[45] The story he is to tell will be one of decline and reformation, in which he perceives five distinct alterations:

> First, I will intreat of the suffering time of the church, which continued from the apostles' age about three hundred years.
> Secondly, of the flourishing time of the church, which lasted other three hundred years.
> Thirdly, of the declining or backsliding time of the church, which com-

prehendeth other three hundred years, until the loosing out of Satan, which was about the thousandth year after the nativity of Christ. During which space of time, the church, although in ambition and pride it was much altered from the simple sincerity of the primitive time, yet, in outward profession of doctrine and religion, it was something tolerable, and had some face of a church; notwithstanding some corruption of doctrine, with superstition and hypocrisy, was then also crept in. And yet in comparison of that which followed after, it might seem, as I said, something sufferable.

Fourthly, followed the time of Antichrist, and loosing of Satan, or desolation of the church, whose full swinge containeth the space of four hundred years. In which time both doctrine and sincerity of life were utterly, almost, extinguished; namely, in the chief heads and rulers of this west church, through the means of the Roman bishops, especially counting from Gregory VII. called Hildebrand, Innocent III., and the friars which with him crept in, till the time of John Wickliff and John Huss, during four hundred years.

Fifthly and lastly, after this time of Antichrist reigning in the church by violence and tyranny, followeth the reformation and purging of the church of God, wherein Antichrist beginneth to be revealed, and to appear in his colour, and his antichristian doctrine to be detected, the number of his church decreasing, and the number of the true church increasing. The durance of which time hath continued hitherto about the space of two hundred and fourscore years; and how long it shall continue more, the Lord and Governor of all times, he only knoweth. For in these five diversities and alterations of times, I suppose the whole course of the church may well be comprised.[46]

Here is Foxe's history in small, in his own words, and here can be seen the great parenthesis that exists in the age of the Church. Its importance can best be perceived by contrast. In Augustine's scheme of six millennia, the Church was to occupy a single temporal space, a sixth and last millennium of illusory duration. In Foxe's history, the time of the Church is fractured and schismatized with difference. It now occupies more than one millennium, and its time can be further subdivided into five periods. The very fact that Foxe can discern these distinctions in his image of the Church's history indicates the dynamism of the conception. Medieval history did not claim that essential unity had obtained throughout all time, but only since Christ, in the age of the Church, for unless time is one time, without alteration, the Church cannot be one Church, without alteration. Foxe finds ages within the last age, and churches within the Church; his subject is the "diversities and alterations of times." Only the earliest and latest churches represent the true Church. Between these is a mutation of Christianity into something unrecognizable as such. The structure attempts a unity of the present with a distant past, the primitive Church, combined with a rejection of the immediate past, the great parenthesis that divides the two ages of truth.

The elegance of the scheme is that it justifies revolution while denying innovation. It is the only possible conceptual structure of time that can tolerate a complete rejection of the past while maintaining substantialism, that is, the belief that anything is permanently true. The past has become innovative (I still use the word in its pejorative sense), while the present revolution is original. The structure is antihistorical, in the sense that it invalidates the truth content of the past, for the union with the early Church is not historical (the links of time have been cut), but textual and typological: it is to be found within the verbal text of the one remaining icon, the Bible, and through the typological identifications that can be derived from the text. The Bible stands alone, deified, removed from the context of history, the communal memory. Whatever mystical participation it may seek in the distant past, Foxe's history is uncompromising in its rejection of the remembered common past that reaches continuously into the present.

The introduction continues with an exhaustive declaration of difference, of separation doctrinal and separation temporal:

> ... the church of Rome, as it is now governed with this manner of title, jurisdiction, life, and institution of doctrine, never descended from the primitive age of the apostles, or from their succession.... In witness whereof we have the old acts and histories of ancient time to give testimony with us, wherein we have sufficient matter for us to shew that the same form, usage, and institution of this our present reformed church, are not the beginning of any new church of our own, but the renewing of the old ancient church of Christ; and that they are not any swerving from the church of Rome, but rather a reducing to the church of Rome. Whereas contrary, the church of Rome which now is, is nothing but a swerving of the church of Rome which then was, as partly is declared, and more shall appear, Christ willing, hereafter.[47]

Much, much more does appear "hereafter." The wedges are driven between the primitive/reformed church and the present church of Rome in every possible area of doctrine and practice; again and again, throughout this massive polemical introduction, are repeated, almost as a ritual response, the words of the desubstantiation of time: "which now is . . . which then was."

Given the vehemency of this prefatory schema, what is surprising about the history itself is its "medievalness." The *Acts and Monuments* is a history devoted to past forms of historiography; its form seems even slavishly imitative and traditional. Its models are Eusebius's *Historia ecclesiastica* (book 1), Bede's *Historia ecclesiastica gentis Anglorum* (most of book 2), the medieval English chronicle of Roger of Wendover, Matthew Paris (and others), and the later version of Polydor Vergil (books 3 and 4). Books five to twelve are Foxe's own history, from Wyclif to his own present, compiled from his own original sources, but this section bears a marked resemblance to Eusebius in form: its subject is the persecution of the Church, martyrdom, and it resolves with the delivery of the Church by a good monarch, Foxe's Elizabeth corresponding to Eusebius's Constantine. Foxe's history is a tour through the tradition, through all the modes and models of Christian historiography that precede it (Augustine is omitted from these, although always understood to be in the background, because Foxe is not formulating universal history, but Church history; therefore, Eusebius is the obligatory model). But Foxe's devotion to his models is deceptive. Within the traditional structure he proposes radical ideas that often negate and reverse the intentions of his originals. Foxe inhabits the old forms in order to take account of them, to discount them. Like one of Machiavelli's usurping princes, he diligently conforms to the customs and usages of the old rulers in order to erase their claim to the loyalty of their subjects, and to legitimize his own squatters' rights in the halls of the past.

The first book of the *Acts and Monuments* is in structure a very close précis of Eusebius. It preserves, without criticism, Eusebius's story of Tiberius's conversion and of Christ's near adoption as a god by the senate,[48] and follows Eusebius in presenting Constantine as the most virtuous of kings and savior of the Church.[49] Yet this exaltation of Constantine also provides Foxe with an opportunity to present Valla's refutation of the Donation of Constantine[50] and the first type in his continuing argument that it is the secular monarch, not the pope, whom God desires to rule and regulate his Church; as Foxe puts it a few pages later, "The king God's vicar within his own kingdom."[51]

There are also many traces of skepticism within this perfect miniature replica of Eusebius. Foxe delights in conflict among the sources when the legitimacy of the papal succession is involved.[52] He disputes the miracles of Saint Alban (imported here out of Bede, because Foxe wants to include Alban within the primitive martyrs), because he discerns parts of the tale to be anachronistic:

> The rest that followeth of this story in the narration of Bede, as of drying up the river, as Alban went to the place of his execution; then, of making a well-spring in the top of the hill; and of the falling out of the eyes of him that did behead him; with such other prodigious miracles mentioned in his story, because they seem more legend-like than truth-like, also because I see no great profit nor necessity in the relation thereof, I leave them to the free judgement of the reader, to think of them as cause shall move him.
>
> The like estimation I have of the long story, wherein is written at large a fabulous discourse of all the doings and miracles of St. Alban, taken out of the library of St. Alban's, compiled (as therein is said) by a certain pagan, who, as he saith, afterward went to Rome, there to be baptized. But, because in the beginning or prologue of the said book . . . the writer maketh mention of the ruinous walls of the town of Verolamium (which walls were then falling down for age, at the writing of the said book, as he saith), thereby it seemeth this story to be written a great while after the martyrdom of Alban, either by a Briton, or by an Englishman. If he were a Briton, how then did the Latin translator take it out of the English tongue, as in the prologue he himself doth testify? Or if he were an Englishman, how then did he go up to Rome for baptism, being a pagan, when he might have been baptized among the christian Britons more near at home?
>
> But among all other evidences and declarations sufficient to disprove this legendary story of St. Alban, nothing maketh more against it, than the very story itself: as where he bringeth in the head of the holy martyr to speak unto the people after it was smitten off from the body; also where he bringeth in the angels going up and coming down in a pillar of fire . . . with other such-like monkish miracles and gross fables, wherewith these abbey-monks were wont in times past to deceive the church of God, and to beguile the whole world for their own advantage.[53]

I reproduce this passage at such length because it perfectly typifies the tone of Foxe's textual criticism. There are assertions here that would be inconceivable in a medieval history: the doubting of sources, not only of the "long story," but also of the great Bede; the distinction between the "legend-like" and the "truth-like"; and the disproving of the author's date by the anachronism of the walls of Verolamium. This last point is crucial for an understanding of Foxe and the new history. Valla had used the accusation of anachronism as one of his chief tools in invalidating the Donation of Constantine.[54] Luther, a great admirer of Valla, had attacked a canon attributed to Anacletus[55] and the *Pseudo-Dionysius*[56] on the same grounds. The very idea of anachronism as a negative quality, as an absence of context within a narrow chronological period, demands, as a presupposition, that history must be conceived of as dynamic, fissured, full of "diversities and alterations."

Foxe's use of Eusebius is in marked contrast to his use of Bede and the English chronicle tradition. He has no real disagreements with Eusebius, who provides him with the source of his first three hundred years, the time of the primitive and faithful Church. This period is structured around "the Ten First Persecutions in the Primitive Church, with the variety of their Torments."[57] Its rhetorical purpose within the larger structure of the *Acts and Monuments* is to establish, from the most ancient and authoritative source, that persecution is the sign of the true Church. Martyrdom is the inevitable result of a true profession. With the commencement of the second book begins the history of the apostate medieval Church, and here also Foxe's use of his sources becomes antagonistic. He will use the structure of Bede and of Roger of Wendover/Matthew Paris, but within that structure he will reverse the ideological content of his sources. The way is prepared by the conclusion of book one, in which Foxe puts forward his interpretation of Revelation. The beast with the seven heads, bearing the whore of Babylon, is the city of Rome (the identification Augustine so carefully avoided). The power of continuing for forty-two months (Foxe reckons each month to stand for seven, "a sabbath of," years) indicates the period between Christ's death and the last year of the persecution of Maxentius or of Licinius, 294 years. It is at this point, in the year 324 according to Foxe, that the great serpent, Satan, is bound, and he will be released a thousand years from then, in the time of Wyclif and Hus. Thus, the binding of Satan only describes a constraint on the active, violent persecution of the Church. During the thousand years, Satan can still deceive and corrupt the Church, but he cannot destroy it by putting its members to death. In the times in which Satan is loosed, the Church must suffer martyrdom; in the times in which he is bound, it will suffer hypocrisy and apostasy, as the outward show of faith is rewarded by the world, and the worldly assume it. During the first period of Satan's raging, his agent was the Roman

Empire, the beast with seven heads; in his second raging, after the thousand years, his agent will be the great whore of Babylon who rides upon the back of the beast, the apostate Roman church.[58] Books two to four chronicle the creation of that great whore.

Foxe's schema allows little rest for the Church. The binding of Satan is no longer identified with the age of the Church, a perfect millennium. The Church, indeed, is to be the agent of the second loosing. Nor can the Church experience both peace and holiness: holiness can only be had in the moment of burning, of torture, in the extremity of martyrdom; deprived of martyrdom, the Church falls from holiness. What small period of relatively holy peace Foxe allows to the Church begins with the accession of Constantine and lasts until the papacy (or, as Foxe would prefer, episcopacy) of Gregory the Great. With Boniface III, put in power by the murdering usurper emperor Phocas, began the Roman supremacy, and with it the inexorable decline of the Church.[59] What follows adheres closely to the chronicle tradition, while destroying all that the tradition asserts: Foxe is skeptical of relics, believing them all to be fraudulent,[60] asserts miracles to be lies[61] (those attributed to Thomas Becket he believes to have been Satanic),[62] and secular princes to be the true vicars of Christ, with power to govern the Church, each in his own kingdom.[63] He is vehement in his hatred of "monkery"[64] (the monks murdered King John).[65] Foxe even hates Saint Francis.[66] The pope, of course, represents the extremity of evil: he sows disunity in the Church;[67] his church is the opposite of Christ's;[68] he should be appointed only by the emperor.[69] Of individual popes, Sylvester was a sorcerer,[70] and Hildebrand attempted to poison the emperor and cast the Sacraments into the fire because they would not prophesy to him.[71] Throughout all, Foxe takes great pains to prove the originality of those doctrines he supports: resistance against the pope,[72] priestly marriage;[73] and the novelty of those he condemns: transubstantiation,[74] purgatory (invented, according to Foxe, by Thomas Aquinas),[75] priestly celibacy.[76] Most subversive of all, Foxe locates the true Church of the Middle Ages in the heretical sects: the Waldenses (who were, he writes, persecuted by antichrist)[77] and the Albigensians.[78] This is the world of the medieval historians turned upside down, a refutation of the past in its entirety. The references I have given are really superfluous; they are only scattered examples. Foxe's overturning of the Catholic past is so densely and vehemently argued that a reader need only open him at random and read a few pages to come across examples of all of these areas of his attack on tradition.

One way to gauge the radicalness of Foxe's history, is to make an inventory of his heroes and villains. First the villains: the list includes, of course, all popes after Gregory, but also Saint Francis, Thomas Becket,[79] and Thomas More.[80] More he regards as a defender of heresy and persecutor of true Christians. He is also particularly incensed by More's jokes, particularly those made on the scaffold, which he recounts in detail. Foxe cannot abide levity in the face of death, for it threatens his whole idea of martyrdom. A martyr must be both solemn and right. Those who die well for the pope's cause he cannot bear, and he must prove that they are not true martyrs. As with Becket, he believes More's condemnation and execution to have been just and (he wishes Becket's had been the same) judicially correct. Nothing can threaten Foxe's world so much as a man who not only showed the courage and steadfastness of the martyr in the service of antichrist, but was also so ironic as to turn his own execution into a comedy show. More did not take his condemnation seriously, and would not have taken Foxe seriously:

> . . . as by nature he was endued with a great wit, so the same again was so mingled . . . with taunting and mocking . . . that he thought nothing to be well spoken, except that he administered some mock in the communication. . . . Also, even when he should lay down his neck on the block, he, having a great grey beard, stroked out his beard, and said to the hangman, "I pray you let me lay my beard over the block, lest you should cut it;" thus with a mock he ended his life. . . . perhaps in the pope's kingdom they [More and Fisher] may go down for martyrs, in whose cause they died; but certes in Christ's kingdom their cause will not stand, howsoever they stand themselves."[81]

These are only a few of the more surprising names on the list of the excoriated. Queen Mary is of course Foxe's arch-villainess, and the Holy Roman emperors he naturally supports in their struggles against the papacy, but condemns when they submit to or support the pope.

On the other side of the tally, Foxe's heroes of the faith of course included Wyclif, Hus, Luther, Edward VII, Elizabeth, and all of the Protestant martyrs of whose fates he writes. But there are also a few surprises on the list. Pico della Mirandola he praises for his criticism of the pope and regards as a true Protestant.[82] Chaucer he declares to have been "a right Wicklevian," and regards all his works to be anti-papist allegories.[83] Saint Bridget of Sweden receives honorable mention for her denunciations of the papal court, as does Saint Catherine of Siena.[84] The most surprising figures in Foxe's canon of Protestant worthies, however, are Anne Boleyn, Henry VIII, and Thomas Cromwell. A good report of the first two was something of a necessity for Foxe; he could not speak any ill of the parents of his sovereign and patron, Elizabeth, but Foxe far transcends the need to give no offense. He presents Anne Boleyn as no less than a Protestant martyr of the greatest spirituality. The accusations against her of adultery and incest are discounted, and the whole process against her is put down to a popish plot, conceived because of her advocacy of the true faith:

> But in this act of parliament [that of 1536 annulling Anne's marriage] did lie, no doubt some great mystery, which here I will not stand to discuss, but only that it may be suspected some secret practising of the papists here not to be lacking, considering what a mighty stop she was to their purposes and proceedings, and on the contrary side, what a strong bulwark she was for the maintenance of Christ's gospel, and sincere religion, which they in no case could abide.[85]

The only name Foxe can supply for these conspirators is that of Stephen Gardiner, bishop of Winchester, who was not even in the country at the time, being at the French court on Henry's embassy. Whether Anne was wholly or partly guilty or innocent of the charges brought against her, the instigators of the action against her were Henry himself and his agent, Cromwell. The papist conspiracy is a nonsensical solution to a problem of nonsensical contradiction: how can Foxe, having made Queen Anne into one of his martyrs, also praise and make faultless the man who had her executed, who remained an orthodox medieval Catholic in all but allegiance to the pope, who was far more strenuous in his persecution of Protestants than of papists (although he killed both), who required adherence to the six articles, "the bloody whip with the six strings"? To Foxe, Henry is always faultless and a man of true religion. His concern over his first marriage was entirely moral and sincere.[86] He receives full credit for breaking with Rome, but in all his actions against the Protestant cause he was always deceived, as was Cromwell, by the plots of papist advisors.[87] Even Cromwell's fall from favor and execution were not the result of any will of Henry's, but of "the malicious craft and policy of divers," divers Romanists of course. It is to Foxe in particular, and to the official school of Elizabethan history in general (in which Shakespeare's *Henry VIII* can be included), that we owe the strange and contrary image of Henry that has persisted even into modern histories: a man swayed by the whim of every advisor, a weak sensualist, driven by appetites rather than politics, whose *Assertio* was ghost-written by More and Fisher (two of those conspiratorial papist advisors); none of which can be reconciled with the man who chose, used up, and threw away the most able statesmen and the most advantageous queens with furious ease and speed, who began maneuvering for his divorce from Catherine when Anne Boleyn was only seven years old.

Foxe's treatment of Henry's reign demonstrates his partiality; it is deliberately revisionist, a rereading and rewriting of history as blatantly careless of truth as Winston Smith's in Orwell's *1984*. Except that Foxe is not really subverting the tradition here, but rather subverting the living memory of events within his own memory and the memories of his contemporaries. Foxe is beginning that tradition, telling the story as the political and religious powers will have it told. At the intersection point at which time is lost and becomes only the story of lost time, Foxe is inventing that story, developing the rhetoric of division. And the third section of the *Acts and Monuments*, the tale of the new martyrdom since the loosing out of Satan, takes place at that intersection point at which the past is becoming lost, becoming text. This section commences with book five, and a dissertation on the loosing out of Satan. From the twentieth chapter of Revelation Foxe derives a structure of the time of the Church, corresponding to the structure of his own history:

> First, The being abroad of Satan to deceive the world.
>   Secondly, The binding up of him.
>   Thirdly, The loosing out of him again, after a thousand years consummate, for a time.[88]

Foxe seems to have some confusion as to whether he is including the thirty years of Christ's life (Foxe evidently ignores the three Passovers of John's gospel, and believes Christ's ministry to have taken place in a single year) and, unless I misunderstand him, gives two different and unreconciled dates for the loosing of Satan: 1324 and 1294.[89] The loosing occurs around the time of Wyclif, when the

Church begins to rediscover the true simplicity of faith, and the church of antichrist reacts by persecuting the true believers. The last eight books are to be the story of these latter-day persecutions, most of which occur within Foxe's lifetime. In structure this last section is united with the first, which was the church history of Eusebius, both ending with the intervention of a God-appointed monarch. Foxe's descriptions of the persecutions, also modeled on those of Eusebius, are hideously explicit and affecting, and these constituted the main attraction of the history—the Book of the Martyrs—for Foxe's contemporaries. Like Eusebius, Foxe attributes miraculous fortitude, even complete freedom from pain, to his subjects, redeeming the squalid deaths by divine intervention (as in Eusebius, the extremity of martyrdom becomes the breaking through into the presence of God, a validating sign to the great mass of believers who do not see face to face): "The report goeth among some that were there present, and saw him burn, that his body in burning did shine in the eyes of them that stood by, as bright and white as new-tried silver, as I am informed by some which were there and did behold the sight."[90] But Foxe's attempts at transcendence are too strained to be convincing (note the repeated emphasis on the witnesses in the last quotation, as though Foxe must enforce some credibility by repeated incantation). Consider this account of the burning of John Hooper, bishop of Gloucester, one of the Marian martyrs:

> In the time of which fire, even as at the first flame, he prayed, saying mildly and not very loud (but as one without pains), "O Jesus, the Son of David, have mercy upon me, and receive my soul!" After the second was spent, he did wipe both his eyes with his hands, and beholding the people, he said with an indifferent loud voice, "For God's love, good people, let me have more fire!" And all the while his nether parts did burn: for the faggots were so few, that the flame did not burn strongly at his upper parts.
>
> The third fire was kindled within a while after, which was more extreme than the other two. . . . In the which fire he prayed with somewhat a loud voice, "Lord Jesus, have mercy upon me; Lord Jesus, have mercy upon me: Lord Jesus, receive my spirit!" And these were the last words he was heard to utter. But when he was black in the mouth, and his tongue swollen, that he could not speak, yet his lips went till they were shrunk to the gums: and he knocked his breast with his hands, until one of his arms fell off, and then knocked still with the other, what time the fat, water, and blood, dropped out at his fingers' ends, until by renewing of the fire his strength was gone,

and his hand did cleave fast, in knocking, to the iron upon his breast. So immediately, bowing forwards, he yielded up his spirit.[91]

What Foxe here takes for miraculous deliverance from pain, for the immediate vision of God, could more easily be interpreted as unimaginable agony and despair. Are repeated screams of "Jesus, have mercy upon me!" and convulsive beating of the breast and working of the lips signs of faith, or merely an uncontrollable flailing, the cries of "Jesus" as futile and hopeless an invocation as a dying soldier's screams for "Mother!"? Foxe's rhetoric, in this and all the other executions he describes, seeks to beatify the filthy grotesqueries of death (". . . saying mildly . . . as one without pains . . . he yielded up his spirit"), to make mere dying into holy dying, in imitation of Eusebius's accounts of the primitive Church.

There are, however, important distinctions between the persecutions described by Eusebius and those described by Foxe: the early Church was persecuted by the pagans, the Protestants by those who professed to orthodox and Catholic Christianity; the accession of Constantine reconciled the pagan empire to the Christian faith, that of Elizabeth made the doctrinal division between Christians into a permanent and national division, England more a nation-state, further from the continent than it had been, and its church a creature of that nation-state, further from the continent of a single Catholic Christianity. This is a situation that cannot be tolerated in Foxe's rhetoric, and is responsible for the tangible strain on almost every page of the *Acts and Monuments*. Foxe must declare the world to be the opposite of its appearances, must prove the great majority of those who profess Christianity to be no Christians and to have no connection whatsoever to the Church of Christ, must show the late persecutions to be the oppression of the holy by the church of antichrist (against which he provokes all the hatred he can; Eusebius is entirely lacking in any similar invective against the pagans). His is a rhetoric against the declared and obvious in the world. He mentions no persecutions of Romanists by the Protestants, nor the drowning of Anabaptists (in parody of immersion baptism) by the German Lutherans. In only one place does any impartiality occur in Foxe's carefully selected executions, and then only because he cannot avoid reporting the event. The impartiality belonged to Henry VIII rather than to Foxe. In 1540 the king had three Protestants and three papists executed at Smithfield on the same day, the Protestants burned for heresy and the papists hanged, drawn, and quartered for treason. Protestant and papist were even drawn in the same hurdle together, and Foxe could not very well report the martyrdom of the former without some mention of the latter (although the accompanying woodcut shows only the burning of the Protestants). The passage indicates something of the horror of doubt that lies under Foxe's clear divisions:

> The same time and day, and in the same place, where and when these three above mentioned did suffer, three others also were executed, though not for the same cause, but rather the contrary, for denying the king's supremacy; whose names were Powel, Fetherstone, and Abel: which spectacle so happening upon one day, in two so contrary parts or factions, brought the people into a marvelous admiration and doubt of their religion, which part to follow and take; as might so well happen amongst ignorant and simple people, seeing two contrary parts so to suffer, the one for popery, the other against popery, both at one time. Insomuch that a certain stranger being there present the same time, and seeing three on the one side, and three on the other side to suffer, said in these words, "Deus bone! quomodo hic vivunt gentes? hic suspenduntur papistae, illic comburuntur antipapistae [Good God! how can anyone live in this place? here they hang papists, there they burn antipapists]." But to remove and take away all doubt hereafter from posterity . . .[92]

And Foxe's next few pages are devoted to clearing up the matter, to explaining the difference between Protestant martyrdom and the execution of papist criminals, how the first is the result of evil council and papist plots swaying the king, the second of his decided good judgment and justice. Behind all this is the terrible possibility that Catholics are equal martyrs, equally sincere, equally courageous in faith, or, even worse, that both sides are equally deluded, die equally for nothing. Baronius's *Roman Martyrology*[93] is the mirror image of Foxe's. The medieval syncretism was dead. The new history, both Protestant and Catholic, had exclusion as its aim.

A broad question remains, prompted by Luther's praise of Valla and by Foxe's praise both of Valla and of Pico della Mirandola: What is the connection between the new Protestant sense of the past and that of the Renaissance? I have concentrated on a particular tributary of historical revolution, one that flows from Wyclif to Hus, then to Luther, then to a whole school of Protestant historians for whom Foxe is the exemplar. Within the same period of time, another revolution had taken place, a second fissure in the medieval unity of time. It begins, as Theodore E. Mommsen has shown, with Petrarch.[94]

The Middle Ages had used metaphors of light and darkness to distinguish historical epochs. The age of the Church, the sixth millennium of the Augustinian system, was an age of enlightenment, in which the world had received the "true light that enlightens every man" (John 1:9 [RSV]), the light that shines in the darkness (John 1:5). The ages of darkness were those before the Incarnation, the pagan epochs of the *preparatio*.[95] Petrarch himself uses just this sense of the metaphor: he pities Cicero for dying before "the end of the darkness and the night of error," before "the dawn of the true light."[96] Medieval classicism was a subspecies of medieval syncretism; it sought to bring all things under the Gospel, even the pagan past. The mission was, however, not merely one of religious duty; it also involved aesthetic devotion. In Petrarch this aesthetic devotion to the Roman past developed into the nexus of his intellectual life; he remained devout, but Christian devotion was not what fired his imagination.

The *preparatio* increased in importance until it became his major love and occupation; the Christian became secondary. Mommsen places the change (it is one of degree, not kind, a subtle transgression) during the composition of *De viris illustribus* and particularly during Petrarch's visit to Rome in 1341. His letters of the period reveal a schism between their Christian statements and their fascination with pagan, not Christian, Rome.[97] The result was a resolution to restrict *De viris illustribus*, originally conceived of as a history of illustrious men throughout time, to the period from Romulus to Titus. After this, Petrarch writes in a letter of 1359, the world entered into an age of *tenebrae*, of darkness.[98] As Mommsen goes on to prove, exhaustively and conclusively, Petrarch meant by his dark age a definite historical demarcation, one in which the world slipped from a Golden Age into a time of worthlessness, of dregs, of barbarity.[99] All history became, to him, "the praise of Rome,"[100] and his hope was for a new age in which something of the Roman past might be revived.[101] Here was a new, twofold division of time: a Golden Age of the radiant past and a present darkness. Flavio Biondo, in his *Decades*,[102] declared the revival, the *renascito*, Petrarch had looked for, and divided history into three distinct periods: the classical, the middle (or dark) ages, and the modern. The *Decades* is the logical systematization of Petrarch's historical statements (as is the *Acts and Monuments* of Luther's), and Biondo is perhaps the most succinct and obvious exponent of the new periodization. Guicciardini, Bracciolini, Valla, Bruni, Sacchi, Macchiavelli— all of the humanist historians explicitly or implicitly believe in Petrarch's and Biondo's sense of time. As Wilcox has pointed out,[103] the Renaissance was defined by the self-conscious division from its own past, and this is what distinguishes it from other supposed renaissances, the Byzantine or that of the twelfth century. All of the humanist thinkers regarded themselves as participants in a great revival of classical learning and art *after* an age of darkness. Witness Vasari:

> Insomma per questa e molte altre cagioni si vede quanto gia fusse al tempo di Constantino venuta al basso la scultura, e con essa insieme l'altre arti migliori. E se alcuna cosa mancava all'ultima rovina loro, venne loro data compintamente dal partirsi Constantino di Roma per andare a porre la sede dell'imperio in Bisanzio. . . . Erano per l'infinito diluvio de'mali ch'avevano cacciato al disotto ed affogata la misera Italia, non solamente rovinate quelle che veramente fabbriche chiamar si potevano, ma, quello che importava piu, spento affatto tutto il numero degli artefici; quando, come Dio volle, nacque nella citta di Fiorenza l'anno 1240 per dar i primi lumi all'arte della pittura, Giovanni cognominato Cimabue, della nobil famiglia in que'tempi de'Cimabui.

> (To sum up, it is clear that for these and various other reasons by the time of Constantine sculpture had already fallen into decline, together with the other fine arts. And if anything were needed to complete their ruin it was provided decisively when Constantine left Rome to establish the capital of the Empire at Byzantium. . . . The flood of misfortunes which continuously swept over and submerged the unhappy country of Italy not only destroyed everything worthy to be called a building but also, and this was of far greater consequence, completely wiped out the artists who lived there. Eventually, however, by God's providence, Giovanni Cimabue, who was destined to take the first steps in restoring the art of painting to its earlier stature, was born in the city of Florence, in the year 1240.)[104]

The structure is one of parenthesis, of the immediate past made empty and waste, a prison from which the modern must escape by reconstituting the ancient past.

With the rejection of the medieval past (our own term "medieval" is the invention of that rejection) went a rejection of the *translatio imperii.* The motive for this has been imputed to political struggles between the Italian cities and the empire, and connected with the development of mercantilism.[105] What is certain is that the rejection of the *translatio imperii* occurs simultaneously with the invention of the dark age: Petrarch denied to Charlemagne his imperial titles, and regarded the Franks and Germans as bastard parodies of the Roman Empire.[106] Both Dante and Petrarch looked for a *renovatio Romae*, but Petrarch's conception involved the restoration of the imperium to Rome: an Italian, not a German, empire.[107] In this can be seen Petrarch's political motive in contrast to Dante's idea of imperial unity in *De monarchia*. In a sense the whole world of the *De monarchia* is far removed from Petrarch's, as is its conception of the *renovatio imperii*, but in one aspect the two are united: the idea of a *renovatio imperii*, however conceived, necessarily destroys the idea of the *translatio imperii*. Whatever requires restoration cannot be truly itself. Dante's very insistence on the unity of time prefigures Petrarch's division of it, and in another generation Leonardo Bruni would have rejected the imperial allegiance to the extent of insisting that Florence owed its foundation to the Roman republic and not to the *imperium*.[108]

At this stage, from this foundation of particulars, it is possible to make some general observations about the interrelationship of Renaissance and Reformation historiography. First, the Renaissance does not precede the Reformation but is contemporaneous with it; the decisive works of Wyclif and of Petrarch straddle the center of the fourteenth century. Because of problems dating Wyclif's works, and because the ideas of both Wyclif and Petrarch develop over the course of several works, an attempt to give specific dates would be absurd, but Wyclif's invention of the age of antichrist and rejection of the papal continuity of the apostolic authority and tradition follow Petrarch's invention of the dark ages and rejection of the *translatio imperii* only by two to three decades. The Renaissance was not a secular movement, nor was the Reformation antihumanistic. Renaissance humanism was in many instances as concerned with the early Church and with doctrine as was the Reformation.[109] Luther felt himself indebted to Valla for the idea of the primitive Church.[110] In fact Luther owed a tremendous intellectual debt not only to Valla's attack on the Donation of Constantine, but also to Bartolomeo Sacchi's history of the popes. Sacchi and Valla are as important to Luther as Wyclif and Hus. Even Luther's "Freedom of the Will" debate with Erasmus—a debate of specific theological issues more attributable to Erasmus's Catholicism than to his humanism, and not per se an attack on humanism—confirms Luther's acceptance of the humanistic terms of the debate and Erasmus's recognition of the importance of doctrine. The Renaissance was not neopaganism but rather a revolt against the institutions and aesthetics of what it would name the medieval world. Religiously, this rebellion against the medieval past could lead to the continuance of Catholic devotion, as it did with Petrarch himself and with More and Erasmus, or to a condemnation of the papacy and a reinvention of the primitive Church, as with Valla, Sacchi, Pico della Mirandola, and the Reformers proper, or to a sharp distinction between the sacred and the profane, as with Bacon and Raleigh,[111] or even to a new cyclical (disguised and muted) paganism.[112] The Renaissance attacked not Christianity but the correlatives of Christianity, the institutions and icons that provided the cultural matrix of faith. A perceived loss of faith was one possible consequence, but a tremendously rare one; a perceived renewal or continuation of faith, but with a different set of cultural correlatives, was another, and one so common as to be almost universal.

Nowhere are the Reformation and the Renaissance so alike as in their critical methodology. The method requires two presuppositions: that history is dynamic, and that early or "original" documents have more value than later ones; the second proposition is clearly dependent on the first, for if history is not dynamic no distinction between early and late would be perceivable. Here, for instance, is Valla on the Donation of Constantine:

> Quid, quod multo est absurdius, capitne rerum natura, ut quis de Constantinopoli loqueretur tamquam una patriarchalium sedium, quae nondum esset nec patriarchalis, nec sedes, nec urbs Christiana, nec sic nominata, nec condita, nec ad condendum destinata? Quippe privelegium concessum est triduo quod Constantinus esset effectus Christianus, cum Byzantium adhuc erat, non Constantinopolis.

> (How in the world—this is much more absurd, and impossible in the nature of things—could one speak of Constantinople as one of the patriarchal sees, when it was not yet a patriarchate, nor a see, nor a Christian city, nor named Constantinople, nor founded, nor planned! For the "privilege" was granted, so it says, the third day after Constantine became a

Christian; when as yet Byzantium, not Constantinople, occupied that site.)[113]

This fragment is typical of Valla's method; it consists in revealing anachronism in its pejorative sense, and presupposes that history exhibits change, mutation, difference. That this sense did not occur before Petrarch is proved by the fact that no one had previously practiced such anachronism-criticism. Petrarch had exposed two contemporary forgeries at the request of the Emperor Charles IV.[114] Forgeries had been immune to such criticism before Petrarch because the historical ideology had not permitted it. It was, in a sense, forbidden to notice that there had been a time when Constantinople was not. Petrarch's project of defining the true texts of the classics by stripping them of later emendations involved a revolution by its very formulation of the idea of "later." Humanism went behind the traditional texts to the original texts, behind the Latin to the Greek, behind the medieval Church to the primitive Church, behind Jerome's Vulgate to the Greek and Hebrew originals.

And this new idea of linguistic and textual time, of contrasting chronism and anachronism,[115] necessitated or was necessitated by a new idea of text, of the text as self-interpreting (so Luther asserted of the Bible), as the author's mind itself standing out in perspective relief from the background of tradition. It is an attitude that seems to originate, as do so many others, with Petrarch,[116] and it depends very much on the availability of texts—to be compared, criticized, used as guides and refuters of tradition. Paracelsus could not have burned the works of Galen had he not possessed them, nor could the Reformation have advocated a biblical instead of a traditional basis of faith until the Bible could be placed in every church. These are the attitudes of print-culture and somewhat precede the development of printing; Petrarch's whole career depended on his access to libraries, Wyclif's on the efficiency of factory-like methods of scribal transcription that allowed the dissemination of his works.[117] The Reformation was very conscious of its dependence on print; Foxe enthuses over printing as God's great revelation to the latter age, the means by which his gospel will be disseminated:

In following the course and order of years, we find this aforesaid year of our Lord 1450, to be famous and memorable, for the divine and miracu-

lous invention of printing. . . . Notwithstanding, what man soever was the
instrument, without all doubt God himself was the ordainer and disposer
thereof; no otherwise than he was of the gift of tongues, and that for a singular purpose. And well may this gift of printing be resembled to the gift of
tongues: for like as God then spake with many tongues, and yet all that
would not turn the Jews; so now, when the Holy Ghost speaketh to the
adversaries in innumerable sorts of books, yet they will not be converted,
nor turn to the gospel.[118]

It is interesting that Foxe does not see printing as an instrument of
conversion, but one of alienation; he rejoices that his enemies will not
repent, that printing will be an instrument in making absolute that
nonrepentance. Every aspect of history is made to serve the rhetoric
of division. Also, the comparison between the pentecostal tongues and
the printing press is an extreme demiracularization, a ridding from the
world of whatever sense remained of the direct presence of God.

This is an appropriate juncture at which to note that the critical
method was rarely an impartial tool, but was almost always the ser-
vant of some partisan cause. Valla attacked the Donation of Constantine at the request of King Alfonzo I of Naples for any attack on the
papacy's claim to the Papal States. Nor is the critical method necessarily disconsonant with credulity. Foxe only doubts those miracles
that advance papal claims and "monkery"; he accepts uncritically the
natural prodigies of the medieval chronicles, the hideous accusations
in them against the Jews, and even reports signs and wonders of his
own, signifying God's and nature's revulsion against Mary's impend-
ing marriage to Philip of Spain.[119]

In all of these aspects the Renaissance and the Reformation are
entirely in agreement; if any difference can be found, it is only in
emphasis, or in the inevitable factions within movements: Luther felt
much more in concert with Lorenzo Valla than he did with Zwingli.
The two revolutions are intertwined, almost inseparable. The Renaissance can claim to be the more comprehensive: there are parts of it
that had nothing to do with the religious reform (although relatively
few parts), whereas all aspects of the Reformation were bound up with
humanism. The two movements began independently, with Petrarch
and Wyclif respectively, and both sprang from the rhetoric of a single
writer, yet became shadows of each other. Both sprang from a rejection of the immediate past, of the *translatio imperii* and of the papal
church, and this may account for their similarities of sympathy and

method. The similarity is even more striking when the new histories of both are considered. The new shape of time involved a great parenthesis driven into time between the distant past—the classical age or the primitive Church—and the present. This dark or middle age, this age of antichrist, must be rejected in favor of a recreation of the distant and ideal past: the golden age, the apostolic age. Bacon said of Luther that his project was, "to awake all antiquity, and to call former times to his succors to make a party against the present time; so that the ancient authors, both of divinity and of humanity, which had long slept in libraries began generally to be read and revolved."[120] With such a statement from such a witness, the popular idea of the opposition of Renaissance and Reformation becomes utterly untenable.

The Renaissance and Reformation precipitated a historical revolution (and it is essentially a single revolution) so profound that it reversed the Western perception of the past within a single generation, from a perception of unity to one of division and difference, from a stillness to a dynamic motion. New ideas had superseded the old, and with supersession came the perception of motion. A person of the early fourteenth century could experience complete intellectual communion with the past; the body of belief had been (it was believed) the same from the foundation of human society; even the Incarnation was not really new, but prefigured in the *preparatio*. A person of the seventeenth century (like one of the twentieth) felt that all who had come before had lived in ignorance. Only the primitive and classical were valued, but they did not touch the present.

4

Histories of a New World

Ancora che per la invida natura degli uomini sia sempre stato pericoloso il trovare modi ed ordini nuovi, quanto il cercare acque e terre incognite.

(The invidious nature of mankind has always made the pursuit of new methods and systems as perilous as the search after unknown seas and lands.)

—NICCOLÒ MACHIAVELLI, *Discourses*

The Atlantic is a Lethean stream, in our passage over which we have had an opportunity to forget the Old World and its institutions.

—HENRY DAVID THOREAU, "Walking"

In the preceding chapters I have attempted to describe the course of two historical revolutions and the systems (which are also, considered in terms of the historical thought they succeeded and destroyed, antisystems or countersystems) that these two revolutions established. My account of the initial Christian historiographical revolution is necessarily partial, as a full description would require an account of the classical pagan historiographies that form the basis, the ground from which the new edifice could rise. A revolution can only be comprehended in the context of what it rebels against, as motion can only be measured relatively, against the medium in which a given body moves.

My second historiographical revolution, that of the Reformation and Renaissance considered as a single, although compound and complex, movement, can be seen in full contrast to its medieval ground. Only in this context does its structure, described in chapter 3 in detail, become evident. That structure is in all aspects the opposite of its medieval antecedent: dynamic instead of static, exclusionary rather than syncretic, defining itself by difference rather than by unity, hostile rather than sympathetic to the Christian Rome of Church and empire, to the city of men and the peace of Babylon. When seen in this contrast, the intellectual and cognitive violence of this revolution stands out in hurtling perspective from the flat background of the medieval past, and assaults the viewer's eye with its velocity and dimensionality along the line of sight. For the new history is above all about movement, the sense of supersession that divides the present from the past, and makes change the arbiter of time.

A sense of time is fundamental to human thought to the extent that the past must be invoked in order to establish any present ideology, even one that involves a discounting of the past. All ideologies are fundamentally descriptions not of a present state, but of a past history. For instance, Marx's description of the inevitable and desirable structure of a future worker's revolution is predicated not primarily on an analysis of present economic relations, but on a doctrine of history. Without a historical narrative of class struggle and reification, Marx's structural analysis of economic relations cannot be defined. The phenomenal world is little more than the idea of its own past.

Granted this importance of history to human thought, what emerges with disturbing clarity from both the initial Christian historiographical revolution and from the second revolution of the Renaissance and Reformation is that in neither case did a new sense of history dictate a new ideology. The opposite is true. New ideology, or theology, demanded the formulation, by an inexorable process of logical progression, of a new idea of the past. Once formed, the new history in turn justified the new theology. History is a literary structure whose literariness must always be denied; its grip on the imagination and on the whole perceived structure of the world is so great that its human origin, its createdness, cannot be acknowledged. The idea of history is one of the most fundamental determinants of social imagination, and the particular historical revolution that emerges from the

Reformation and Renaissance is one of the most fundamental determinants of modernity. Aesthetics, technology, cosmology, economy,
theology, self-fashioning individualism—all of the aspects of the
period have received considerable attention, but less so the concomitant historical revolution. This we are reluctant to describe, for its idea
of time is our own, and more fundamental and persistent than any
other cultural inheritance.

For a study of Protestant historiography beyond Foxe (and the new
history contains within it always the idea of going beyond), the historians of the New World offer decided advantages. Theirs is the furthest
removal from the medieval world and Church, a removal not only
from Catholicism, but also from previous waves of in-their-eyes
imperfect reformation, and an ideological, intellectual, and imaginative removal reinforced by three thousand miles of physical removal.
To alienate the past was the object of the migration, to find a history-
less world in which the literary model of the New Testament Church
could be reconstituted as flesh:

> Let the matter and forme of your Churches be such as were in the Primitive
> Times (before Antichrists Kingdome prevailed) plainly poynted out by
> Christ and his Apostles, in most of their Epistles, to be neither Nationall
> nor Provinciall, but gathered together in Covenant of such a number as
> might ordinarily meete together in one place, and built of such living stones
> as outwardly appeare Saints by calling.[1]

The parenthesis in this passage, containing antichrist's kingdom, is
the same parenthesis the New England Puritan would put in time
itself, the waste time of antichrist's kingdom being removed from the
reckoning, leaving only the present and the "Primitive Times." In
such a venture, histories should neither be necessary nor useful; the
Bible, the textual model, should stand alone as the atemporal template
by which to create the social world. Yet histories—perhaps contradictory and tormented histories, describing the past in order to deny it—
did emerge. If a utopian New World needs, as Hawthorne observed, a
cemetery and a prison, it also needs a history.

William Bradford commences his *Of Plimoth Plantation*[2] with a statement of the same parenthetical history (stated orthographically with
the same literal parenthesis as used by Edward Johnson in the quota-

tion above) that I have proposed as the essential constitutive paradigm
of the Protestant historical model:

> It is well knowne unto the godly and judicious, how ever since the first
> breaking out of the light of the gospell in our Honourable Nation of
> England, (which was the first of nations whom the Lord adorned ther with,
> affter that grosse darknes of popery which had covered & overspred the
> Christian worled,) what warrs & opposissions ever since, Satan hath raised,
> maintained, and continued against the Saincts, from time to time, in one
> sorte or other. Some times by bloody death and cruell torments; other
> whiles imprisonments, banishments, & other hard usages; as being loath his
> kingdom should goe downe, the trueth prevaile, and the churches of God
> reverte to their anciente puritie, and recover their primative order, libertie,
> & bewtie.[3]

Here are all the elements of Protestant and Puritan history. The typical sentence has for its subject a supernatural agent, either God or
Satan, who enacts the active verb of the historical event. All earthly
history is merely the emanation of movers in the unseen world. There
are no human, or political, or material causes; all discrete and discern-
able disturbances in the flow of the world begin, like the Book of Job,
in the court of heaven; and this metahistory, this history beyond and
behind the history of merely material and observable events, subsumes
mere history into itself, and into the rhetoric that creates it and can be
its only expression. The form this metahistory takes is that of a rejection of the recent and historical past, the "grosse darknes of popery,"
which is relegated to the temporal and printed parenthesis, in favor of
an identification with a mythical and textual past, one not continuous
with the present, a recovery of an "anciente puritie," a "primative
order, libertie, & bewtie."

This is the supersessive history of the revolutionary, the model of
time invented by Petrarch and by Luther; by the time of Bradford's
writing it has been rehearsed to the point of formula. There is, however, something new in Bradford's expression of it: something is badly
wrong with the time after the parenthesis. Instead of, or coexistent
with, the "anciente puritie" and "primative order" revived, there are
"warrs & opposissions ... bloody death and cruell torments ...
imprisonments, banishments, & other hard usages." These date not
from the Catholic persecutions of the Protestant martyrs, the subject
of Foxe's history, but from that time "since the first breaking out of

the light of the gospell in our Honourable Nation of England." These events of course include the defiling of Wyclif's corpse and the Marian persecutions, and to this extent conform to Luther's and Foxe's model of the new martyrdom. But there is much more implied here. The passage continues:

> But when he [Satan] could not prevaile by these means, against the maine trueths of the gospell, but that they began to take rooting in many places, being watered with the blooud of the martires, and blessed from heaven with a gracious encrease; he then begane to take him to his anciente stratagemes, used of old against the first Christians. That when by the bloody & barbarous persecutions of the Heathen Emperours, he could not stoppe & subuerte the course of the gospell, but that it speedily overspred with a wounderfull celeritie the then best known parts of the world, he then begane to sow errours, heresies, and wounderfull dissention amongst the professours them selves, (working upon their pride & ambition, with other corrupte passions incidente to all mortall men, yea to the saints them selves in some measure,) by which wofull effects followed; as not only bitter contentions & hartburnings, schismes, with other horrible confusions, but Satan tooke occasion & advantage therby to foyst in a number of vile ceremoneys, with many unproffitable cannons & decrees, which have since been as snares to many poore & peaceable souls even to this day. So as in the anciente times, the persecutions by the heathen and their Emperours, was not greater then of the Christians one against other; the Arians & other their complices against the orthodoxe & true Christians. . . .
>
> The like methode Satan hath seemed to hold in these later times, since the trueth begane to springe & spread after the great defection made by Antichrist, the man of sine.[4]

I quote Bradford at such length here because these passages demonstrate clearly one of the distinctives of the Puritan colonial histories and of the second phase of Protestant historiography. Here is Luther's conception of the dynamic decline of the Christian religion, the anti-traditionalist doctrine of the gradual rush from unmediated and pure spiritual truth into the degraded simulacra: the "vile ceremoneys," the "unproffitable cannons & decrees." Within this process is the remnant of the true Church, which only and ever exists in a state of persecuted opposition to apostasy. All of this Bradford shares with Luther and with Foxe, but whereas they described the decline of the primitive Church into papist corruption, and the persecution of the true and suffering remnant at the hands of that antichristian papacy, Bradford

describes a decline of the Reformation itself. Whereas Luther and
Foxe used the rhetoric of martyrdom developed by Eusebius in
response to pagan persecutions and transferred it to the persecution of
Protestant Christians by Catholic Christians (thereby declaring the
larger portion of Christendom to be not Christendom), Bradford
transfers this rhetoric again to the persecution of Protestants by other
Protestants. After the great parenthesis, he reconstitutes not only the
primitive Church, but also the decline of the primitive Church. The
first fall into heresy, before renewal came, took a millennium and a
half; the second decline only one generation.

What is at work here is the continued principle of supersession. The
first generation of Reformation histories declared the truth to have
been reestablished after the decline into antichrist's kingdom; for
Bradford, the new truth has once again repeated the structure of
decline and needs yet another reestablishment, another revolution.
And establishment is the conscious goal of every revolutionary society.
The truth of which the past was ignorant must now be made eternal
and universal, for what the supersessor most fears is continued and
unstoppable supersession. For Bradford, the first supersession of the
English Reformation has already declined into sin, and demands a
new supersession that will this time persevere. And as supersession
recurs, the necessarily concomitant sense of separation from that
which is superseded, both the past and the larger culture, becomes
both infinite and infinitely subtle:

> For to let pass the infinite examples in sundrie nations and severall places
> of the world, and instance in our owne, when as that old serpente could
> not prevaile by those firie flames & other his cruell tragedies, which he by
> his instruments put in ure every wher in the days of queene Mary & before,
> he then begane an other kind of warre, & went more closly to worke; not
> only to oppuggen, but even to ruinate & destroy the kingdom of Christ, by
> more secrete & subtile means, by kindling the flames of contention and
> sowing the seeds of discorde & bitter enmitie amongst the proffessors &
> seeming reformed them selves.[5]

The subject, the enactor, of this single, long sentence is Satan, the
"old serpente"; the earthly history of the Marian persecution is insubstantial compared to the power of its supernatural cause and is relegated to prepositional phrases within a subordinate clause: "which he
by his instruments put in ure every wher in the days of queene Mary
& before." And the subject even of this subordinate clause, becalmed

away from the main race of the sentence, is Satan; the earthly history is reduced to "instruments" and "days." The grammar embodies the hierarchy of the epistemology: the supernatural world, or, rather, the literary rhetoric of the Bible, which creates the supernatural world in the imagination, exerts almost absolute precedence over any merely empirical knowledge.

So, failing by instruments of outright persecution—and these instruments, it must be remembered, are professed Christians, although the rhetoric of the new martyrdom has by this stage excluded the fact by silence more powerful than the previous, shrill identifications of the larger Christendom with antichrist—Satan has gone "more closly to worke," by "secrete & subtile means," which must inevitably be more painstakingly discerned by the faithful. These means are contentions, "discords & bitter enmitie amongst the professors & seeming reformed them selves." Although the point may be oversubtle, Bradford's separation of "them" from "selves" is appropriate to his theme; within the division of Christian from antichristian is the new division of the presumably truly reformed from the "seeming reformed," the new persecutors who provide the impetus for Bradford's history of separation and removal.

The historical matter, the earthly events, of Bradford's narrative, following the metahistorical introduction, are persecution, exile, and contention, all introduced by citation of the history's *auctour* and point of departure, John Foxe:

> Mr. Foxe recordeth how that beside those worthy martires & confessors which were burned in queene Marys days & otherwise tormented, *Many (both studients & others) fled out of the land, to the number of 800. And became severall congregations. At Wesell, Frankford, Bassill, Emden, Markpurge, Strausborugh, & Geneva, & c.* Amongst whom (but especially those at Frankford) begane that bitter warr of contention & persecution aboute the ceremonies, & servise-booke, and other popish and antichristian stuffe, the plague of England to this day, which are like the high-places in Israell, which the prophets cried out against, & were their ruine; which the better parte sought, according to the puritie of the gospell, to roote out and utterly to abandon. And the other parte (under veiled pretences) for their ouwn ends & advancments, sought as stifly to continue, maintaine, & defend.[6]

The exile in search of purity divides into a bitter contention, into "*severall congregations*," both multitudinous and severed, whose only resolution can be a further exile of the more pure from the less pure, the truly reformed from the seeming reformed. The problem of the seeming reformed is their rootedness in history, in "ignorante & super-stitious times," their "popish trash, which have no ground in the word of God, but are relikes of that man of sine."[7] Just as the allegation of novelty was, to the medieval world, an automatic condemnation, so, to Bradford and his like, is the allegation of tradition, of connection to the past, for the past is a human invention, but the "puritie of the gospell," "the word of God"—this latter by now no longer designating the incarnate Christ of John's Gospel but instead transmuted entirely into the Bible—is atemporal, a literary model having no connection to history but descended immediately from the divine.[8] Within this understanding of history, history itself becomes the subject and act of idolatry, and the goal of those who seek the purity of the gospel must be a complete severance from the historical past. Thus each new exile, each new schism is a further separation not only from the (increasingly smaller) larger community, but also from the past and history itself, the shedding of a series of carapaces of what was once quick time.

Bradford relates how, after their return from the Marian exile, the English Reformers established their church under James I in the man-ner of Edward VI's precedent, which differed from the Reformed churches in Scotland, France, the Netherlands, and Geneva, "*whose reformation is cut, or shapen much nerer the first Christian churches, as it was used in the Apostles times,*"[9] that is, nearer to the textual model of the primitive Church, before the intervening parenthesis of history and tradition, antichrist's realm. In response to this imperfec-tion of reformation, a group of "godly & zealous" in "the North parts"

> shooke of this yoake of antichristian bondage, and as the Lords free people, joyned them selves (by a covenant of the Lord) into a church estate, in the fellowship of the gospell, to walke in all his wayes, made known, or to be made known unto them, according to their best endeaours, whatsoever it should cost them, the Lord assisting them. And that it cost them something this ensewing historie will declare.[10]

After their secession from the established and imperfect church, "These people became 2. distincte bodys or churches." One of the churches, which chose John Smith as its pastor, declined, despite its early promise: "But these afterwards falling into some errours in the Low Countries, ther (for the most part) buried them selves, & their names."[11] It is the other church, led by Richard Clifton, John Robinson, William Brewster, and Bradford himself, "which must be the subjecte of our discourse." This group removed to Amsterdam in order to avoid persecution by the established church and its civil sponsor, and there yet another division occurred:

> And when they had lived at Amsterdam aboute a year, Mr. Robinson, their pastor, and some others of best discerning, seeing how Mr. John Smith and his companie was allready fallen in to contention with the church that was ther before them, & no means they could use would doe any good to cure the same, and also that the flames of contention were like to breake out in that anciente church it self (as aftterwards lamentably came to pass); which things they prudently fore-seeing, thought it was best to remove, before they were any way engaged with the same."[12]

The company removed to Leiden, but here they were beset by many outside sources of temptation, due to the liberty of the place, which particularly affected the children, "So that they saw their posteritie would be in danger to degenerate & be corrupted."[13] The answer is yet another removal, this time to Virginia, but this decision is attended by schisms and near schisms with those members who will not go, and those who prefer Guyana as a destination.

This is the structure of Bradford's first book: an odyssey of removal after removal by an ever-diminishing company. Each body identified with the pure and primitive Church falls into a corruption of contention, and the impure, those still in bondage to the continent of Europe and the continent of the past, are repeatedly sloughed off by the pure who attain, or attempt to attain, the ahistorical and atemporal form of the primitive Church: "they came as near the primative patterne of the first churches, as any other church of these later times have done."[14] In all of this there is a terrible contradiction.

Sacvan Bercovitch has remarked that Bradford's history differs from those of the Massachusetts Bay Puritans in its Separatist ideas of church establishment.[15] The distinction is a true one although it can be overstated. Bradford did not see Plymouth as a divinely ordained holy state, a fusion of pure church and pure society; it was, like Amsterdam, Leiden, the Guyana his fellowship had considered, or the Virginia they had actually tried to get to when accident or providence drove them north, an accident rather than an essential of church history, another place where freedom might be found. Yet although

Bradford does not seek establishment with the same typological intensity as the historians I will consider later, establishment is nevertheless
his goal. He writes of "the holy discipline & outward regimente of the
kingdom of Christ, by which these holy doctrines should be conserved,
& true pietie maintained amongst the saints & people of God."[16] His
fellowship of Separatists "joyned them selves . . . into a church
estate."[17] Although Bradford's goal is not that of the holy state, it is
that of the holy congregation. The goal of the enterprise is to pass
through the stages of imperfect reformation in order finally to attain
to "the primative patterne of the first churches." Here is the tension:
a model of history and faith in which a principle of dynamic supersession has as its end establishment. At each stage of their odyssey of
removals in the Old World and the New, Bradford's church anticipated that it had at last found the civic liberty in which establishment
of a true church would be possible, only to be unhoused[18] by dissension and, at Leiden, the falling away of the youth from excess of liberty. Against this history of perpetual and possibly unstoppable exile
within exile, separation within separation, Bradford closes his first
book, the book of the wanderings, with landfall in a New World, and
the concrete image of establishment:

> On the 15 *of Desemr:* they wayed anchor to goe to the place they had dis-
> covered . . . and they arrived safe in this harbor. And after wards tooke
> better view of the place, and resolved wher to pitch their dwelling; and the
> 25. *day* begane to erect the first house for comone use to receive them and
> their goods.[19]

A great deal of attention has been paid to the problematic structural
and thematic relations between the two books of Bradford's history.[20]
The second book abandons the chapter divisions of the first, and with
them the idea of an ordered narrative. Rather, Bradford writes, he will
give the rest of his history,

> (if God give me life, & opportunity) . . . for brevitis sake . . . by way of
> *annalls,* noteing only the heads of principall things, and passages as they
> fell in order of time, and may seeme to be profitable to know, or to make
> use of.[21]

The very form seems already, in this introductory note, to betoken
some loss of control over the interpretation of earthly events. Bradford

began the first book (which concludes with the Pilgrims landcoming
and erection of the first common building in 1620) in 1630, at the
beginning of the much larger Puritan migration to the Massachusetts
Bay Colony, and completed it soon after. The second book he began
in 1646, with the reach of twenty-six years' knowledge of the Pilgrims'
condition in the New World, and wrote all the annals up to his present
date retrospectively.[22] Yet it is the first book that has the form of con-
fident and completed history; in adopting the form of annals for the
second book, Bradford deprives himself of the providential and teleological historian's perspective from a position at the end of time,
from which the final apotheosis can be anticipated. Eusebius formed
his narrative of the Church's history during the age of persecutions
from the perspective of Constantine's peace; Augustine descried the
six ages of the world from the last of these, at the completion of God's
revelation; Luther depicted the age of antichrist from the perspective
of one who had lived to see the rebirth of the true doctrine, which
immediately preceded the end of the world; Foxe narrated the persecutions of the Protestants from the perspective of the Elizabethan settlement, his description of which he models on Eusebius's Constantinian settlement—in each case the historian creates a narrative of
conflict and resolution, and the resolution must be declared permanent, the plot of history essentially complete. The temporal *topos* from
which the past is reinterpreted (in any interpretation strong enough to
create or alter historical consciousness) is the frozen cusp of the last
wave of the vicissitudes of human history, poised at the static point of
equilibrium before it falls beyond the world.

The first book of Bradford's history concludes at this point, with the
remnant of the true Church established in the New World, poised for
some unstated and ineffable apotheosis into the kingdom of God.
With the first paragraph of the second book, that apotheosis is lost.
Bradford has cast away the ability to create a narrative of completion,
to set form against time and mutation. In the annal, time itself
becomes the only ordering function. He will write only, "(if God give
me life, & opportunity)," and he chooses the new form—"*annalls*,
noteing only the heads of principall things, and passages as they fell in
order of time"—"for brevitis sake." "If God give me life, & opportunity" gives an ambiguous shadow to "for brevitis sake": it is not only
the brevity of the entries themselves, but the brevity of life, which
motivates the brevity of writing, and not the brevity of Bradford's life

alone, but the universal condition of brevity, of life among the things
that will not stay. And the expansion of reference given to "for brevitis
sake" similarly expands the reference of "as they fell in order of time,"
particularly as it is contiguous to the erection of the first house for
common use, with which the first book ends. Whatever the second
book may be about, and that will remain uncertain to the end, it cannot be about establishment (the form will not allow it); the strange
interplay of seemingly simple terms in its first paragraph transforms it
from the outset into an ambiguous and doubtful enterprise, mirroring
the ambiguous and doubtful enterprise it describes, likely to be about
the fall of many things under the order or rule of time. The language
displays in its ambiguous self-unraveling the sensibility of a sojourner
who has come to the promised land and found himself still a
sojourner.

In his description of the Pilgrims' first landfall, Bradford exhibits the
hatred of the land that is so characteristic of all the New England Puritans[23] (so different from the demi-eden of the Old Dominion settlers),
and so prevalent that Samuel Sewall's relatively late (1697) description
of the beauties of Plum Island is held up as a rarity:[24]

> Besides, what could they see but a hideous & desolate wildernes, full of wild
> beasts & willd men? . . . Nether could they, as it were, goe up to the tope
> of Pisgah, to vew from this willdernes a more goodly cuntrie to feed their
> hops; for which way soever they turnd their eys (save upward to the heav-
> ens) they could have little solace or content in respecte of any outward
> objects. . . . If they looked behind them, ther was the mighty ocean which
> they had passed, and was now as a maine barr & goulfe to seperate them
> from all the civill parts of the world.[25]

The sense of exile here is acute and poignant. The search for a
home, which led the group to leave Amsterdam and Leiden, has led
instead to a perpetual wilderness sojourn. Not even as a Pisgah top,
from which to view the promise, will the land serve.

The spirit of this landfall is appropriate, for it is the land, its winters
and its hardships, that will figure so prominently as the subject of the
second book. Bradford writes, as did Eusebius at the beginning of
church history, an account of the pragmata, the earthly dealings of the
Church in exile. The Pilgrims did achieve their goal: to find a place of
freedom and isolation in which to recreate the pattern of the primitive

churches without oppression from the civil authority or the lax and corrupting influence of a libertarian and libertine society. But the establishment of the true Church is followed by time, by the expansion of that mutating duration which is the absence of God. This is not to say that the second book does not tell of God's providences, but even these perceived providences are pragmata, physical remedies to physical trials. Bradford becomes, in Bercovitch's words, a secular historian, a recorder of "the fate of a wholly temporal venture."[26] When he does write about the spiritual state of the plantation, it is to write, for instance, about Mr. Charles Chansey, who began a baptism controversy in the Plymouth church: "he holding it ought only to be done by diping, and putting the whole body under water, and that sprinkling was unlawfull,"[27] unlawful because it was "an humane invention." Here again is the automatic disqualifier to the Reformation historical mind, as novelty was to the medieval. The Plymouth church answered very sensibly that immersion was certainly lawful, but very inconvenient "in this could countrie." All attempts at compromise failed and the result was another schism, another unhouseling of Chansey and his followers to Scituate, "Sityate."

Immediately following this account in the annal for 1641 is this:

> Also about these times, now that catle & other things begane greatly to fall from their former rates, and persons begane to fall into more straits, and many being allready gone from them . . . and many other, & stille some dropping away daly, and some at this time, and many more unsetled, it did greatly weaken the place, and by reason of the straitnes and barrennes of the place, it sett the thoughts of many upon removeall.[28]

This in turn is immediately succeeded by the annal for 1642, which contains the great catastrophe of the latter part of the history:

> Marvilous it may be to see and consider how some kind of wickednes did grow & breake forth here, in a land wher the same was so much witnesed against, and so narrowly looked unto, & severly punished when it was knowne; as in no place more, or so much, that I have known or heard of; insomuch as they have been somewhat censured, even by moderate and good men, for their severitie in punishments. And yet all this could not suppress the breaking out of sundrie notorious sins . . . espetially drunkennes and unclainnes; not only incontinencie betweene persons unmar-

ied, for which many both men & women have been punished sharply
enough, but some maried persons allso. But that which is worse, even
sodomie and bugerie, (things fearfull to name,) have broak forth in this
land.[29]

Bradford ponders the causes of this outbreak of uncleanness in the
plantation, and suggests several possibilities: first, that the devil carries
a greater spite against the churches of Christ in the New World because
of their strenuous efforts for purity, but:

> I would rather thinke thus, then that Satane hath more power in these hea-
> then lands, as som have thought, then in more Christian nations, espetially
> over Gods servants in them.
>   2. An other reason may be, that it may be in this case as it is with waters
> when their streames are stopped or dammed up, when they gett passage
> they flow with more violence, and make more noys and disturbance, then
> when they are suffered to rune quietly in their owne chanels. So wikednes
> being here more stopped by strict laws, and the same more nerly looked
> unto, so as it cannot rune in a comone road of liberty as it would, and is
> inclined, it searches every wher, and at last breaks out wher it getts vente.[30]

What Bradford is proposing here is a theory of equilibrium that
makes the establishment of the true church-state or even the true congregation impossible. Temptation increases proportionately with the
approach to perfection; sin increases in pressure proportionately with
its suppression. Establishment is the creator of dissolution, and the
history concludes with a series of dissolving images: the baptismal
schism, the removal of the population to other settlements, the great
outbreak of wickedness to which Bradford devotes so much detail. He
fills pages with the disquisitions on the sins of sodomy and buggery,
and their punishments, provided by the ministers of Massachusetts
Bay in response to questions from Plymouth (these documents not
only emphasize the outbreak of wickedness, but also indicate the
increasing intellectual subservience of the Plymouth colony to Massachusetts Bay), and concludes with the case of Thomas Granger, who,

> being aboute 16. or 17. years of age . . . was this year detected of buggery
> (and indicted for the same) with a mare, a cowe, tow goats, five sheep, 2.
> calves, and a turkey. Horrible it is to mention, but the truth of the historie
> requires it. . . . A very sade spectakle it was; for first the mare, and then the

cowe, the rest of the lesser catle, were kild before his face, according to the law, Levit: 20. 15. and then he him selfe was executed.[31]

The passage is indicative of Bradford's essential doubleness of perspective: a sorrow and horror both at the crime and at its punishment, but also an agreement with the punishment inflicted.

What remains of Bradford's history after this moral cataclysm is dissolution and pragmata: the death and biography of William Brewster, the decline of the Plymouth church through the repeated removals of portions of its congregation:

> And thus was this poore church left, like an anciente mother, growne olde, and forsaken of her children, (though not in their affections,) yett in regarde of their bodily presence and personall helpfullness. Her anciente members being most of them worne away with death; and these of later time being like children translated into other families, and she like a widow left only to trust in God. Thus she that had made many rich became her selfe poore.[32]

The history ends with inconsequential matters: the squabbling of agents for the colony, the absence of Mr. Edward Winslow in England for four years, "which hath been much to the weakening of this govermente," and then finally two empty entries: "*Anno* 1647. *And Anno* 1648."[33] Thus conclude the "passages as they fell in order of time." The struggle of the historical narrative that attempts to order and overcome time, to see the past and present as pattern and completion, has ended; the narrative voice lapses into inconclusive silence, and only time remains, annals without entries. Finally there is an appendix, a listing of all those who voyaged on the *Mayflower*, a turning back to 1620 to register the dying and the dead. Beyond this last silence lies the private study of Hebrew, "[an] effort to resurrect the literal language of some original perfection."[34]

There is in Bradford from the beginning, but increasing as his chronicle increases in size and in the duration it records, an opposition between providential history and secular annals; between a belief in the possibility of establishing a true church on earth, "the primative patterne of the first churches," and a perceived necessity for continued schism, removal, unhouseling, an equilibrium of establishment and dissolution. Bradford retains an almost medieval sense of separation

between history and God, combined uneasily with an attempt at providential and even, in a few places, typological history. The migration
may, uncertainly, be God's preordained purpose, or it may be only a
self-willed accident in an approach to God, the *telos* of which lies
always and only beyond the world, beyond history.

Bradford's subjunctive climb of a Pisgah that is not there reveals no
promised land, but an emptiness, "(save upward to the heavens)," that
affords no "solace or content in respecte of any outward objects." This
emptiness is for Bradford the very ground of faith, the acutely perceived absence of God in history that witnesses the existence of God
by the very ability of the human soul to perceive that absence. This
ultimately becomes the purpose of the history, implicit from the first
and explicit at the end: to witness the dissolution of its own establishment, to part the tissues that connect earthly history to the divine, and,
leaving the supersessive years to themselves, to seek an original and
eternal perfection whose form is a blank page, an abandonment of the
historical and communal for privacy, Hebrew, and death.

By the time of Bradford's last annals in *Of Plimoth Plantation*, the
Massachusetts Bay Colony had become the keeper of the Reformation
spirit in New England. This was the many the Plymouth church had
made rich while making herself poor, and all of Bradford's comments
about the much larger colony, while they are superficially supportive,
display some tone of unease. The first published work of history in
Massachusetts Bay and in New England is Edward Johnson's *Wonder-
Working Providence of Sions Saviour in New England* (London,
1654), and it is a remarkable document, by comparison with which
Bradford comes to appear a muted convert to Reformation history, a
writer not quite at ease with his own rhetorical mode. Johnson, captain of militia, surveyor of boundaries, and inspector of armaments to
the General Court, has proved particularly repulsive to twentieth-century taste and, to redress the imbalance, has recently been the subject
of eloquent defense.[35] Ursula Brumm excuses Johnson for his lack of
impartiality and critical judgment by pointing out that he is not writing history in the modern sense but rather a record of "wonder-working providence," as his title proclaims:

> Nowhere does it claim to be historical: it intends to be, as the title says, the
> record of God's providence in New England. It is not historical, even in its

external form and character, because it does not mean to be historical in our meaning of the term.[36]

The statement is a true one, but it needs to be made much stronger: no histories were meant to be historical in "our" meaning of the term before the nineteenth century, and even these are becoming increasingly remote in their affinity to a continually shifting present. Johnson is trying to write history within the understanding of his Reformed contemporaries: his is precisely the history of Luther, Foxe, Bale, and Brightman, adapted to the needs and matter of the migration. If modern taste prefers Bradford to Johnson, that is probably due to Bradford's eventual failure in the rhetoric of Reformation history and Johnson's success.

Wonder-Working Providence, from its very first words, exhibits precisely the characteristic elements of Protestant and Puritan history revealed in the first pages of Bradford, but with a new pitch of rhetoric. As with Bradford, no earthly causes are allowable: in each sentence the acting subject is supernatural, the verb an intersecting operation of the supernatural upon the natural:

> When England began to decline in Religion, like lukewarme Laodicea, and instead of purging out Popery, a farther compliance was sought not onely in vaine Idolatrous Ceremonies, but also in prophaning the Sabbath, and by Proclamation throughout their Parish churches, exasperating lewd and prophane persons to celebrate a Sabbath like the Heathen to Venus, Baccus and Ceres; in so much that the multitude of irreligious lascivious and popish affected persons spred the whole land like Grashoppers, in this very time Christ the glorious King of his Churches, raises an Army out of our English Nation, for freeing his people from their long servitude under usurping Prelacy; and because every corner of England was filled with the fury of malignant adversaries, Christ creates a New England to muster up the first of his forces in; Whose low condition, little number, and remotenesse of place made these adversaries triumph, despising this day of small things, but in this hight of their pride the Lord Christ brought sudden, and unexpected destruction upon them. Thus have you a touch of the time when this work began.[37]

Here again is the second decline into idolatry and apostasy: not the decline of the Catholic Church, but of the English Reformation, God's answer to which is a second reformation within the first. As each reformation is by definition not a reformation of the whole Church (reformation can only exist in opposition to that which necessitates reformation: antichrist's church) but only of a faithful remnant, a second reformation must be attended by schism and removal, by unhouseling.

All of these characteristics Johnson's introduction shares with Bradford's, but there are also some notable differences. Whereas Satan is the acting subject of most of Bradford's sentences, "Christ the glorious King of his Churches" is the actor in Johnson's. He "raises an Army out of our English Nation, for freeing his people from their long servitude under usurping Prelacy"; he "creates a New England to muster up the first of his forces in"; he "brought sudden, and unexpected destruction upon [his enemies]." Bradford's concentration is on history as a record of Satan's oppression of God's people; Johnson's is on a history of Christ's militant triumph over his enemies. While the theology of history is identical, the extrapolations from that theology are radically different in effect: Bradford's Church is a suffering remnant, whose duty on earth is to remain faithful while enduring hardships and persecutions at the hands of false Christians, not even papists but "seeming reformed"; Johnson's Church is the remnant militant, whose task is to become mighty despite its low condition and little number, and to crush the enemies of God. For Bradford the mark of the true Church is that it is persecuted; for Johnson that mark is that it persecutes the enemies of God.

Another remarkable departure from Bradford evident in this passage is Johnson's use of typology ("lukewarme Laodicea," "Grashoppers"), which will become more concentrated as his history progresses. Much has been written on Puritan typology, and I do not intend to explicate Johnson's typology as such.[38] Yet some clarification is necessary in order to understand the function of Johnson's types. Christian typology originated in the New Testament writers' reinterpretation of the Old Testament, and thus has apostolic precedent for both Protestant and Catholic.[39] The symbolic vector of early and medieval typology is from the earthly to the spiritual, from Canaan to heaven, from Israel to the Church (considered as an eternal mystery, spiritual Israel), from the rock in the desert to Christ. As time progressed the typological interpretations became fixed, part of a general interpretive tradition.

New typological identification is inherent to periods of reinterpretation; typology is the interpretive tool that can best be used to dismantle the traditional, to open hidden meanings. Inevitably, then, the Reformation was accompanied by a plethora of new typological identifications, and these new identifications linked mainly Old Testament types to current earthly and political events, reversing the earlier direction of fulfillment.[40] Types were for the Reformation what relics were for the medieval Church; they linked the earthly present to the divine past, allowing men in their mundane lives, separated from the divine, to possess a little bit of heaven. Types are of course not physical objects, but literary tropes, held in the mind, and permissible to the Reformation because they are not physical, nor "humane invention," but God's symbols, pieces of the Bible, the only permissible and remaining icon to unite present and past, spilled over into the present world. This spilling over is important to the Protestant idea of the type, and can best be illustrated by a type that is not specifically biblical:

> About this time there fell out a thing worthy of observation, Mr. Winthrop the younger, one of the magistrates, having many books in a chamber where there was corn of divers sorts, had among them one wherein the Greek testament, the psalms and the common prayer were bound together. He found the common prayer eaten with mice, every leaf of it, and not any of the two other touched, nor any other of his books, though there were above a thousand.[41]

To the believer in typology, the biblical types were not mere literary symbols, but were God's symbols in the world, the record of which was embedded within the biblical text. The distinctive of the Reformation typology is the belief that God continues to create antitypes, the fulfillments of biblical types, in contemporary earthly history, and also to create new symbols. This passage from Winthrop exemplifies God's contemporary providential symbolizing. The mice are supernatural agents, sent from God to eat the *Book of Common Prayer* in order to show that it is a "humane invention" and not to be used in services. The New Testament, particularly the Greek original, and the Psalms are God's own books. The symbol is not of human origin, but is created by God; the interpretation is not free, but is determined by God, and so obvious that Winthrop suggests no interpretation; it is, as Luther asserted of the Bible, self-interpreting to any who will read it honestly and in faith.

The type therefore acts as an interpenetration of time and eternity,

in the case of the biblical type an interpenetration of the present with
that original, mythical and literary history of redemption, beyond the
parenthesis of antichrist's kingdom. That the types, God's symbols,
are not merely literary but are still being fulfilled and created by miraculous providence in the physical world provides a new medium of
temporal unity, makes time palindromatic, capable of being read both
forward and backward.[42] This is not the same attempt to hold all time
in a temporal unity that characterizes medieval historiography, but an
attempt to unite two points across a chasm, to declare what is new to
be original, to make a dynamic, onward flow of time a double movement rather than a stasis. This same palindromatic movement of time,
in which the Church of England and Laodicea become symbols of one
another, is reflected in a confusion of tense in Johnson:

> Christ Jesus intending to manifest his Kingly Office toward his Churches
> more fully than ever yet the Sons of men saw, even to the uniting of Jew
> and Gentile Churches in one Faith, begins with our English Nation (whose
> former reformation being vere imperfect) doth now resolve to cast down
> their false foundation of Prelacy, even in the hight of their domineering
> dignity. And therefore in the yeere 1628, he stirres up his servants as the
> Heralds of a King to make this proclamation for Voluntiers, as followeth.
>   *"Oh yes! oh yes! oh yes! All you the people of Christ that are here
> Oppressed, Imprisoned and scurrilously derided, gather yourselves together,
> your Wives and little ones, and answer to your severall Names as you shall
> be shipped for his service, in the Westerne World, and more especially for
> planting the united Collonies of new England; Where you are to attend the
> service of the King of Kings."* [43]

The exhortation beginning *"Oh yes!"* continues for six chapters in
the future and present tenses and in the imperative, and even after the
proclamation of the heralds, the history shifts bewilderingly between
historical, present, and future, tending always to the purely prophetic
voice. This is history as exhortation, and exhortation cannot be conveyed in the past tense.

Already, in these first two pages of the history, there is an extremely
problematic relationship between Johnson's rhetoric and material
events. James I's and Charles I's Declaration Concerning Sports of
1617 and 1633 is hardly adequate to the pagan sabbath to Venus, Bac-
cus, and Ceres that Johnson describes. Nor did Christ literally create
a New England, as though it rose out of the sea, nor did heralds of
Christ proclaim in these words the exhortation to leave for America.

Here is more than partiality; the *Wonder-Working Providence* is a history of metahistory, of the secret workings of God behind the masks and movements of the world, of the substance of history to which political and physical events bear only the relation of accidents. Bradford in his introductory account of the course of the Reformation attempts such a history, but when he comes to the tale of his own church, history and metahistory draw apart; God comes to be perceived without history (in both senses of without) and history without God. Bradford uses types rarely, and when he does so it is in problematic and often negative terms: he invokes a Pisgah that is not there from which there is no promised land to be seen. Johnson, by contrast, translates history into a divine metahistory through a force of rhetoric that consumes the phenomenal world. The history beyond takes absolute precedence, sweeps the preterit past into a future exhortation which can then be experienced retrospectively, as history, by the children of the exhorted. This is history in the court of heaven, history as it should have been and will, in the cycles of God's types, be again.

The exhortation continues with a prophetic history of establishment, an establishment both earthly and apocalyptic:

> It was further proclaimed as followeth: "What, Creature, wilt not know that Christ thy King crusheth with a rod of Iron, the Pompe and Pride of man, and must he like man cast and contrive to take his enemies at advantage? No, of purpose hee causeth such instruments to retreate as hee hath made strong for himselfe; that so his adversaries glorying in the pride of their power, insulting over the little remnant remaining, Christ causeth them to be cast downe suddenly forever, and wee find in stories reported, Earths Princes have passed their Armies at need over Seas and deepe Torrents. Could Caesar so suddenly fetch over fresh forces from Europe to Asia, Pompy to foyle? How much more shall Christ who createth all power, call over this 900 league Ocean at his pleasure, such instruments as he thinks meete to make use of in this place, from whence you are now to depart, but further that you may not delay the Voyage intended, for your full satisfaction, know this is the place where the Lord will create a new Heaven and a new Earth in, new Churches, and a new Commonwealth together."[44]

Here again is Johnson's typical sentence structure: "Christ thy King crusheth with a rod of Iron." The typological significance of this and of the other sentences of which Christ is the subject is very particular, as significant for what it excludes as for what it contains. Johnson's Christ is not the Christ of the Gospels or of the Epistles, but of one book only, the Johannine Apocalypse, in which he appears glorified and terrifying, the judge of the world, the righteous avenger who treads the wine press of the fury of the wrath of God.[45] Even in Revelation, the picture of Christ is complex, and in the symbolic mode of the book this complexity is expressed by an opposition of symbolic representation: the promised Lion of the tribe of Judah is instead a lamb, who appears as though slain; even in the passage of the white rider, Christ's robe is dipped in the blood of his own sacrifice.[46]

All of this complexity is purged from Johnson's imagery; all traces of the Christ of love and mercy are removed, omitted from a Christ who, by means of his attendant images, is entirely identified with the Lord of Hosts of the Old Testament. In a survival of the medieval idea that all history is a source of eternally applicable example (and this idea persisted in the Reformation for all history not included within the parenthesis of antichrist's reign), Caesar is invoked as another type of the militant Christ and his militant Church. The passage concludes with an absolute conflation of apocalypse and earth: Christ promises a new heaven and a new earth in America; it is here the New Jerusalem will descend. Yet this promise is accompanied by the promise of new churches and a new commonwealth, as though these things were no less marvelous. Time and eternity are made one in a rhetoric of religious establishment on earth. But even this supremely emphatic language of establishment has encoded within itself the possibility of failure:

> At your landing see you observe the Rule of his Word, for neither larger nor stricter Commission can hee give by any, and therefore at first filling the Land whither you are sent, with diligence, search out the mind of God both in planting Church and civill Government, but be sure they be distinct, yet agreeing and helping the one to the other; Let the matter and forme of your Churches be such as were in the Primitive Times (before Antichrists Kingdome prevailed) plainly poynted out by Christ and his Apostles, in most of their Epistles, to be neither Nationall nor Provinciall, but gathered together in Covenant of such a number as might ordinarily meete together in one place, and built of such living stones as outwardly appeare Saints by calling ... if you shall slight, despise or contemne [the elders of the churches], hee will soone frustrate your call by taking the most able among you to honour with an everlasting crown; whom you neglected to honour on Earth double as their due, or he will carry them remote from you to more infant Churches. You are not to put them upon anxious Cares for their daily Bread, for assuredly (although it may now seeme strange)

you shall be fed in this Wilderness, whither you are to goe, with the flower of Wheate, and Wine shall be plentifull among you (but be sure you abuse it not). The Doctrines delivered from the Word of God imbrace, and let not Satan delude you by perswading their learned skill is unnecessary, soone then will the Word of God be slighted as translated by such, and you shall be left wildred with strange Revelations of every phantastick brain."[47]

This passage contains all of the tensions and contradictions of Puritan history; these occasion the extremity of Johnson's rhetoric and are occasioned by it. The very terms of commission and exhortation ("see you observe," "with diligence, search out," "be sure") diminish God's immanent control of history. The will of God has become the province of human agents whose success is by no means guaranteed (this is one result of the almost exclusively Old Testament identifications of the typology), charged with a large and strict commission. The churches and the government are to be distinct; Wyclif, Luther, and Foxe had all taught (in reaction to the hated idea of a universal and self-governing Church) that the secular prince should have authority over the Church in his own kingdom, but the Church of England was just such a church as advocated by the Reformers (this further illustrates the position of Puritanism as a reformation within a reformation, a schism within a schism). Yet, despite this separation, the church and government are to be always in agreement, and the government to exist as a means of furthering religious aims, in effect a church-state. The churches are to recreate the pattern of the "Primitive Times" before the antichristian parenthesis; here a palindromatic linking of times is coupled with a rejection of the historical past, represented by the time within the parenthesis and by the continent of the past from which God's commissioned are separating themselves both physically, intellectually, and temporally. The pattern of the primitive Church is "plainly poynted out" in the Bible; it is, as in Luther's conception of biblical, doctrinal truth, self-interpreting. Yet the colonists must follow the teaching of the ministers and elders who have the learning to expound the word. If this discipline of interpretation breaks down, God will carry the faithful "remote from you to more infant Churches"; the result of failing to heed authoritative teaching will be further schism, further removal, repeated supersession; the colonists will become apostate and a new remnant will be chosen. The passage is filled with the rhetoric of establishment, yet "you shall be fed in this wildernesse whither you are to goe." Elsewhere, Johnson describes the colonists as a "Wildernesse-People,"[48] as "wandering Jacobites" whose retreat from the "Desert" and "terrible Wildernesse" is prevented by the sea.[49] His image of God is that of the pillar of cloud.[50] The rhetoric of the temple, of establishment, of the living stones, is drawn always back into the rhetoric of the tabernacle, of exile and wandering and continual process. Finally, there is the choice between obedience to a clerical discipline (in a writer and a culture that hates "usurping prelacy") and being "left wildred with strange Revelations of every phantastick brain."

The rhetoric is an attempt to reconcile and combine within one intellectually homogenous and consensual society contradictory and incompatible ideas. The most important of these antinomies concerns history and the idea of temporal structure. The Reformation established the idea of ideological supersession. For the first time in the history of Europe since the beginning of the Christian era a new thing had been discovered in time, a secret hidden by God to be revealed in the last generation: the form of true Christian worship and doctrine known to the primitive churches. Whereas the medieval image of time had preserved an absolute intellectual identification with the past, the Reformation allowed no such identification to its adherents: for fifteen hundred years all had been deceived, apostate, wrong. The meaning of history reversed itself within a generation. But the new history still retained the essentialist ideals of the old. Any truth must be eternally and unchangingly true, and innovation—the "newfanglednes" that Bradford goes to such lengths to prove is not the reason for the Pilgrims' removal[51]—remains an invention of the devil. The answer to this dilemma was to accuse the past, the time within the parenthesis of antichrist's kingdom, of innovation, of dynamic and supersessive decline from the eternal pattern of the true Church. The Reformation, by contrast, was truly timeless in that it united one eternity with another into changeless unity.

Reformation historiography, therefore, retains, while giving new value to, an ideal of atemporal essentialism, yet simultaneously creates the idea of supersession, movement, and change in history, both by its description of the history of the Church's decline, and by the precedent of its own revolution, which must be declared not to be the innovation it obviously is, but an original and timeless perfection. The repressed material of this mode of historical thought, the terror that motivates the surface of the historical rhetoric, can best be phrased as a question:

If the structure of history is one of change and supersession, and if the principle of revolution, of ideological supersession, is established, then how can the new truth, which must be called original truth, be established; how can further revolution and further supersession be prevented? As the Reformation progressed, the question became acute: Lutheranism, Zwinglianism, Calvinism, Anabaptism—and within each new revolution and schism, hundreds of other removals within individual churches. For each new supersessive group, each new revolution within revolution, the goal of establishment.

The second great antinomy grows naturally from the first, and can also be put most immediately as a question: How can a society that valorizes the individuality of interpretation and faith reach consensus?[52] For lack of consensus, individual and several interpretation, is the inevitable agent of continued supersession. The question of consensus, therefore, cannot be separated from the historical model; it is determined by it and will determine it for the future.

These two antinomies are the motives for Johnson's rhetoric. They condition the violence of its assertion, which, while it attempts to reconcile opposites, makes those opposites more extreme. On the question of consensus he warns:

> ... beware of any love self-conceited Opinion, stopping your ears from hearing the Counsell of an Orthodox Synod.

> ... heale not lightly the wounds that Wolves make, less from their festering Teeth a Gangrine grow.

> Yea, such will be the phantasticall madness of some (if you take not heed) that silly Women laden with diverse lusts, will be had in higher esteeme with them, then those honoured of Christ, indued with power and authority from him to Preach.

> And therefore be sure there be none to hurt or destroy in all his holy Mountaine, and as he hath pressed you for his service, that by passing through the Flouds of Persecution you should be set at liberty, and have power put into your hands, Then let none wrest it from you under pretence of liberty of Conscience. Men of perverse judgements will draw Disciples after them, but let your consciences be pure, and Christs Churches free from all Doctrines that deceive. And all you, who are or shall be shipped for this worke, thinke it not enough that you injoy the truth, but you must hate every false way and know you are called to be faithful Souldiers of Christ, not onely

to assist in building up his Churches, but also in pulling downe the King-dome of Anti-Christ, then sure you are not set up for tollerating times, nor shall any of you be content with this that you are set at liberty, but take up your Armes, and march manfully on till all opposers of Christs Kingly power be abolished.[53]

The tension in these passages, more eloquent and compact than any paraphrase of them could be, is between the principle of individual interpretation and the need to establish a *magisterium* of churchly authority, of conformity in faith and practice. These exhortations are part of the commission given in England by "the Heralds of a King" before the removal. The third passage ("silly Women laden with diverse lusts") is of course meant to be a prophecy of Anne Hutchin-son (this is another example of the problematic relations between Johnson's creation of the past and what, in a modern popular sense, might be stated as the truth of history), the pattern to the point of obsession in the Puritan mind of all those who would initiate further schism and supersession. Johnson's history is a call for *magisterium* and persecution, and the nature of these does not differ from the Cath-olic practices or those of the Anglicans. The *magisterium* to be enforced inside the colony is contained within a militant sense of sep-aration from those without:

Your Magistrates shall not but open the Gates for all sorts. But know, they are Eyes of Restraint set up for Walles and Bulworks, to surround the Sion of God; Oh for Jerusalem her peace, see that you mind it altogether, you know right well that the Churches of Christ have not thrived under the tolerating Government of Holland.[54]

Edward Johnson, surveyor of boundaries, twice in *Wonder-Work-ing Providence* surveys the boundaries of the Christian state and lists those to be hated. I will give these extraordinary catalogs in full, for they convey better than any other passages the picture of the world wrought by adherence to the farthest reaches of the Reformation image of history:

. . . never to make League with any of these seven Sectaries.
  First, the Gortonists, who deny the Humanity of Christ, and most blas-phemously and proudly professe themselves to be personally Christ.
  Secondly, the Papist, who with (almost) equall blasphemy and pride pre-

fer their own Merits and Workes of Supererogation as equall with Christs unvaluable Death, and Sufferings.

Thirdly, the Familist, who depend upon rare Revelations, and forsake the sure revealed Word of Christ.

Fourthly, Seekers, who deny the Churches and Ordinances of Christ.

Fifthly, Antinomians, who deny the Morrall Law to be the Rule of Christ.

Sixtly, Anabaptists, who deny Civill Government to be proved of Christ.

Seventhly, The Prelacy, who will have their own Injunctions submitted unto in the Churches of Christ. These and the like your Civill Censors shall reach unto. . . .

Listen a while, hear what his herauld proclaimes, Babylon is fallen, is fallen, both her Doctrine and Lordly rabble of Popes, Cardinalls, Lordly-Bishops, Friers, Monks, Nuns, Seminary-Priests, Jesuits, Ermites, Pilgrims, Deans, Prebends, Arch-Deacons, Commissaries, Officialls, Proctors, Somners, Singing-men, Choristers, Organist, Bellows-blowers, Vergers, Porters, Sextons, Beads-men, and Bel-ringers and all others who never had name in the Word of God; together with all her false Doctrines, although they may seeme otherwise never so contradictory, as Arians, who deny the God-head of Christ, and Gortenists who deny the Humanity of Christ: Papists, who thinke to merit Heaven by the Workes of the Law, Antinomians, who deny the Law of God altogether as a rule to walke by in the obedience of Faith, and deny good works to be the Fruit of Faith, Arminians, who attribute Gods Election or Reprobation to the Will of Man, and Familists, who forsake the revealed Will of God, and make men depend on strong Revelations, for the knowledge of Gods Electing Love towards them, Conformitants or Formalists, who bring in a forme of worship of their owne, and joyne it with the worship God hath appointed in his Word, Seekers, that deny all manner of worship or Ordinances of Christ Jesus, affirming them to be quite lost, and not to be attained tille new Apostles come.[55]

The intellectual isolation exhibited by these two lists, the tremendous extremity of affect shown by the need to enumerate the ranks of the hated, is extraordinary, even frightening. The world of mind it inhabits is a tiny pocket of orthodoxy, beset from without by the past—the papists, the prelacy, and the formalists; and from within by the future—Antinomians, Quakers, Seekers, all the various Protestant cults spawned by the great impulse toward individual interpretation, interiority, and supersession. The past the Puritans are trying to destroy; the future threatens to destroy them by a continued application of their own principles of transmutation and schism, threatens to make their would-be eternal establishment of the one truth just a transitory wave, just a temporary and random configuration within an ever-expanding intellectual diaspora. *Wonder-Working Providence* represents with frightening clarity the historical consciousness of a culture ground between the upper and the nether millstone, hating both past and future. The principle of the true Christian society, writes Johnson, is enmity:

> Namely, the great enmity betweene that one truth as it is in Jesus, and all other unsound and undeceiveable Doctrines, together with the persons that hold them . . . as sixteene hundred yeares experience will testifie.[56]

In place of Augustine's "We also profit from the peace of Babylon," Johnson sets "Babylon is fallen, is fallen." Instead of the syncretism of the early Christian intellect, he erects a great particularism, "Walles and Bulworks" set against the past, against the possibility of future change, against Europe, the Old World, the larger culture. One of his favorite images is that of the burning glass: God has caused the "dazeling brightnesse of his presence to be contracted in the burning-Glasse of these his peoples zeale."[57] The whole of history is for Johnson the record of the contraction of that light from its diffusion in the larger world to its concentration in the faithful remnant. In the most daring of his typological identifications, the Puritan colony becomes not only God's chosen people but even a new Incarnation (a comparison of the burning-glass passage with the first verses of John's Gospel makes the identification inescapable), a contraction of God's light to a fiery point, attended by heavenly portent, a blazing star of the new nativity: "the blazing Starre (whose motion in the Heavens was from East to West, poynting out to the sons of men the progresse of the glorious Gospell of Christ, the glorious King of his Churches)."[58]

At the point in Johnson's historical argument (and in my description of it) at which separation, the narrowing of the boundaries of intellectual sympathy, seems to have reached a point without dimension, the ultimate focus of that burning glass, another set of contradictions must be taken into account. Enmity expands to salvation. If the Puritans are freed to slaughter the native American "Amelakites" through the Old Testament destruction-of-peoples typology, they also strive for their salvation.[59] The burning glass not only concentrates the light of God in the New World, but also brings the fire of Reformation to Europe both for the destruction of antichrist's kingdom, and for the salvation of all:

> . . . it begins to be left upon many parts of the World with such hot reflection of that burning light, which has fired many places already, the which shall never be quenched till it hath burnt up Babilon Root and Branch.[60]

Destruction and salvation are ever close in Johnson's mind, and their association represents his and his culture's generosity. A culture so cut off in sympathy must find some justification for itself, some avenue for mental expansion. That expansion is found in the future (and Johnson's history, with all its confusion of tenses, is preeminently about the future), in apocalyptic purpose.

I have written of this mode of history's hatred of the future, but that needs to be qualified. What makes Reformation history distinctive is its hatred of the past, in contrast to the temporal and cultural syncretism of medieval history; its hatred of the future it shares with medieval history. But the phrase "hatred of the future" must be refined to "hatred of the futures it does not imagine." Each system of history is enthralled by the future it has projected, its *telos* and apocalypse by which it gives meaning to the whole of time; what makes a society look to its inquisitors is the possibility of futures that go beyond and threaten to invalidate the accepted apocalyptic program. In Johnson's apocalypse can be found the Puritans' reintegration with the world or, rather, its reintegration with themselves.

Johnson's proclamation had promised "this is the place where the Lord will create a new Heaven and a new Earth in, new Churches, and a new Common-wealth together."[61] I have already remarked on the conflation of history and eternity in this passage; as Johnson becomes more eschatologically specific, the conflation becomes more marked:

> Keepe your weapons in continuall readinesse, seeing you are called to fight the Battails of your Lord Christ; who must raigne till hee hath put all his enemies under his Feet, his glorious Victories over Antichrist are at hand, never yet did any Souldier rejoyce in dividing the spoyle after Victory, as all the Souldiers of Christ shall, to see his judgement executed upon the great Whore, and withall the Lambs bride prepared for him, who comes Skipping over and trampling down the great Mountaines of the Earth, whose universall Government will then appeare glorious, when not onely

the Assyrian, Babilonian, Persian, Grecian and Roman Monarchies shall subject themselves unto him, but also all other new upstart Kingdomes, Dukedomes, or what else can be named, shall fall before him; Not that he shall come personally to Regne upon Earth (as some vainly imagine) but his powerfull Presence and Glorious brightnesse of his Gospell both to Jew and Gentile, shall not onely spiritually cause the Churches of Christ to grow beyond number, but also the whole civill Government of people upon Earth shall become his, so that there shall not be any to move the hand, nor dog his tongue against his chosen, And then shall the time be of breaking Speares into Mattocks, and Swords into Sithes; and this to remaine to his last comming, which will be personally to overcome the last enemies of his Saints, even death, which hee will doe by the word of his Mouth, audibly spoken the World throughout.[62]

Augustine had identified the thousand-year reign of Christ with the sixth millennium, the age of the Church, and in this age Satan is bound; Wyclif followed Augustine, but believed that Satan was now loosed to usher in the age of antichrist, the thousand years being over; for Johnson, the thousand year reign of Christ is about to begin, marked by the Puritan errand that will bring about the fall of antichrist, the eradication of Romanism from the world. Only after the fall of antichrist will Satan be bound and the pure Church established.[63]

Johnson's chiliastic vision has the strange effect of making the apocalypse the psychic opposite of its Augustinian medieval antecedent. In Augustine's scheme the return of Christ was made both infinitely near and far by the last millennium. When it did come, it would be an intervention from without, a translation of the Church from the world of the dark glass into the face-to-face, the immediate, supernatural, unimaginable presence of God. For Johnson the coming of Christ is to be figural only, a rule on earth by means of the pure Christian Church. Only after the Church has reigned uncorrupted for a thousand years will Christ personally and supernaturally return. The expected Second Advent and the age of the Church, separate in Augustine, have here become one. It is as if the whole history of the Church, as it had been understood, was only a false beginning, a failure to commence. All must now begin again. But it cannot be argued that Johnson's expected rule of Christ through the Church is not the Second Coming; it unavoidably occupies that place in human expectation. Rather, the Second Advent has become secularized. It is now an earthly reign through human agents ("you are called to fight the Battails of your Lord Christ"), and Christ is not personally present.[64] It becomes therefore a subject of great anxiety, for if Christ's kingdom depends on human agents it is possible that the failure of those agents will cause the failure of the kingdom of God. It is the task specifically of the New England Puritans to bring about that kingdom; it is theirs "to re-build the most glorious edifice of Mount Sion in a Wildernesse."[65] The result of this building in the wilderness is to be the salvation of England ("for Englands sake they are going from England to pray without ceasing for England")[66] and not England only but of all Europe and the world: the New England Puritans are

> but the Porch of his glorious building in hand ... the whole Nation of English shall set upon like Reformation according to the direct Rule of his Word? Assured confidence there is also for all Nations, from the undoubted promise of Christ himselfe.[67]

The passage continues into an enfolding of all nations into Christ's new reformation, into the millennial kingdom of which the Puritans represent only the "forerunners of Christ's Army." This is the generosity of Johnson's vision of history. There is a sense in which the American Puritans are never the real point of God's purpose. This of course increases the sense of exile. They are the forerunners, the example by which the world will attain true reformation and the Antichrist be destroyed. It is their task to be a peculiar people, called out from the world, to maintain the purity of the gospel in embattled but splendid isolation, a shining city, set upon a beacon hill in the New World, to be seen by all.

This vision of universality in the midst of intellectual separation from the past and from the larger world is of course dependent on a belief in the predicted future. The rhetorical system of Johnson's history contains within itself the possibility of its own destruction in all of the movements it condemns, any of which has the potential to warp the future from its predicated course. His descriptions of the hated sectaries sometimes read as a recapitulation of the original decline of the Church.[68] One of the prominent last events in his second book is the death of the first generation.[69] Yet to the very end, Johnson's rhetoric succeeds in combining, although under force, the disparate elements and inherent contradictions of his culture. His second book, more pragmatically historical and less exhortatory and theoretical, is an account of the planting of new towns and new churches, of a continuing expansion of that point of light into universality and apocalypse. Unlike Bradford's *Of Plimoth Plantation, Wonder-Working Providence* succeeds in knitting time to eternity, and in this represents the highest achievement of Puritan historical rhetoric, but the combination is unstable, less a solution than a suspension, sustained by brilliance and heat of an extreme language. It is a momentary achievement.

Writing more than forty years after Johnson, Joshua Scottow, in his *Narrative of the Planting of the Massachusets Colony*, exhibits the contingent futures purged by Johnson's language.[70] The structure perceived by Scottow in the colony's history is ably expressed by the title he gave to the earlier and shorter version of the same work (Boston, 1691): *Old Mens Tears for Their Own Declensions.*[71] After Johnson's onward rush of time, Scottow's historical sensibility seems almost shockingly retrospective. He sings the decline of the errand, and a history that leads him to a "perplexed Labyrinth, of Distracting thoughts of heart." His history is occasioned by

> The Late Series of Divine Dispensations tending not only to the dissolving of the Cement, but to the subverting of the Basis of that Fabrick which the Wonderful Worker hath here so stupendously erected, nor to the Cropping off their Branches, but to the Rooting up of the tender Plant, which the Heavenly Father, here so graciously hath Planted . . . the Aspect of Providence so terribly varying, from what formerly it was wont to be.[72]

In a perfect datum of Cecilia Tichi's thesis, the metaphors of the living building—attempts to create solidity and substance from the insubstantiality of language—declare themselves to be dissolved into the airy nothingness from which they were "so stupendously erected." The sentences of supernatural causality ("the Wonderful Worker hath here so stupendously erected . . . the Heavenly Father, here so graciously hath Planted") are retrospective and contained within subordinate clauses. Indeed the language of these first lines of the history seem to have no syntactical direction or movement at all; they typically consist of a subject, the series of dispensations, lost in a labyrinth of modification, waiting for the delivering verb that, when it comes after long delay, immediately introduces a substitute subject:[73]

> hath put some of the Old Relict Planters, upon . . . serious considerations of what provoking evils we have committed.

This is to be a history of human subjects and their sins; the divine subject exists for Scottow only in the past: "Such was the *Day of Christ's Power*."[74]

Both Bradford and Johnson describe the passing of the first generation among the last things of their histories; Scottow begins with a memoir of the death of the revered John Cotton, the great prophet of the first generation:

> How near *New-England* now is to its breaking, the all-knowing One only knows; but the muteness of this Prophets Doctrine, is with all solemnity and sadness of soul to be Lamented. . . . This Prophet is Dead, *and our Fathers where are they?*[75]

The entire history is written from the perspective of God's silence, but the silence is filled with hieroglyphs, bequeathed by a generation of prophets who knew the interpretation. The typological mode that expressed itself in extreme interpretive assertion in Johnson and in the episode of the mice in Winthrop's journal, an assertion so strong it denied the human act of interpretation and declared itself simple recognition of self-interpreting revelation, in Scottow becomes an obsession with interpretation itself. His is a God-haunted world in the sense that God has become unfamiliar, uncanny in Freud's sense of *unheimlich*, is revealed in apparitions, half-seen, fleeting, in acts and signs that "must symbolize something unseen."[76] Even when Scottow commemorates ("Commemorate" is a recurring word in the *Narrative*) the divine establishment of the colony (which, he protests, too strongly for the good of his own argument, was not a commercial or earthly venture),[77] God's enactments have about them an air of terrible ineffability:

> Infinite Wisdom and Prudence contrived and directed this Mysterious Work of Providence, Divine Courage and Resolution managed it, Superhumane Sedulity and Diligence attended it, and Angelical Swiftness and Dispatch finished it; Its Wheels stirr'd not, but according the HOLY SPIRITS motion in them; yea there was the Involution of a Wheel within a Wheel; God's Ways were a Great Depth, and high above the Eagle or Vulturous Eye; and such its Immensity as mans Cockle-shell is infinitely unable to Emptie this Ocean.[78]

The image of Ezekiel's wheels, which recurs throughout the history,[79] is a strange typological identification, because, even in its original biblical context, it is uninterpretable, representing God self-revealed as unfathomable enigma, enigma within enigma, "a Great Depth, and high above the Eagle or Vulturous eye." The world is strewn with God's signs, but God no longer gives the interpretation. There is laid on mankind an imperative to interpret, but no surety of success. And interpretation is made more difficult by the pastness of God's acts. Scottow posits an original point of unity from which human and divine have diverged and diverge further with the passage of time:

> That the Great God is Departing from us, his Awful Removes demonstrate
> . . . should not this bring to Remembrance, both our Personal Fall in our
> First Fathers, and our Relative Apostacy from our Church-State? First
> Love, and First Works! Our Fathers were Clothed with the Sun, the Apos-
> tolical Discipline and Doctrine were their Crown, the Moon was under
> their Feet, but we are turned topsy turvy, Heads and Heels have changed
> places.[80]

The causes of this fall were inherent in the colony from the beginning:

> . . . a Snake crept forth, which Lay Latent in the Tender Grass . . . the Ser-
> pents subtilty shew'd it self in a Multitudinarism of Questions, started
> under pretence of seeking Light; Error cloath'd it self under disguise of
> Truth by pretext of Magnifying Grace, it was turned into Wantonness.[81]

These two passages make plain what has happened to the colony's perception of history between Johnson and Scottow: it has succumbed to its own model. Scottow's fall is simply a re-creation of the fall of the English Reformation with which Bradford's and Johnson's histories began. This in turn was a recreation of the fall of the Catholic Church, which fall is the basis of Reformation history. Each generation of the Reformation historical model argued its own exception from the overall structure of dynamic decline and supersession. But the structure itself has tremendous and seductive power, tending to overwhelm all who accept it. It is not only a literary and structural model, passed as influence from text to text, although that is the aspect of it that I have traced. The model has also a social dynamic, insepa-rable from its undeniable literariness, and tending always to disinte-gration, to new ideologies, to separation of the present generation from the past and from the larger society. It was forged as a tool of conflict, a way of experiencing history that could be used as a weapon against tradition, against the claims of the past, against establishment, for it pits and privileges the claims of individual interpretation, of new discovery of truths hidden in time, against the need of any ideologically cohesive society for *magisterium*. While proclaiming a centripetal gathering in of all nations into God's one and eternal truth, the Reformation tended always to centrifugation, to "multitudinarism," to the separation of sect from community and present from past.[82]

Ideologies, myths—the verbal structures describing things unseen—tend to take their revenge on the tactile world; the subtleties of the verbal description of God or of history take precedence over the screams of the dying or the smell of burning flesh. Foxe's martyrology is the extended account of just such a revenge. The wonder of the sixty-year span of New England's history that is the subject of Scottow's "perplexed labyrinth" is that it contains neither revolution nor civil war nor the self-devouring horror of the Anabaptists at Münster. The only deaths directly attributable to the inevitable conflict between centripetal *magisterium* and centrifugal supersessionism are those of the four Quakers hanged on Boston Common between 1659 and 1661. This low number of victims was due partly to the habit of exiling ideological recalcitrants, a punishment that, while it was designed to remove "multitudinarism" from the society, actually advanced centrifugation. What the conflict did inevitably produce, however, was an absolute incompatibility between ideological coherence and civic order. What won out in Massachusetts, due to the complex pressures of royal governance among many other causes, was a natural inertia in favor of civic order. As the society grew in social and mercantile complexity, and as schism multiplied, as the model of history and of interpretation dictated it must, then the enforcement of ideological uniformity became inimical to material cooperation.[83] The result was an inevitable and progressive toleration of "multitudinarism."

To return now to the beginning of Scottow's history, the divine dispensations that occasion the work are the Indian wars. The Indians are the inhuman pagan and Satanic enemy, the Amelakites,[84] who are stirred up by God to punish his unfaithful people. Yet the presence of this external enemy is almost a blessing in the history. It shows (perhaps) that God still acts, even in fury, and allows a jeremiadic structure to the *Narrative*. Scottow's real subject, however, is not the dispensation, but the sin that occasioned it, which sin is none other than the history of progressive toleration. His history is one of increasing physical security and prosperity, from which the Indian wars seem a deliverance, coupled with a loss of ideological coherence, of any sense of truth or surety. His condemnation is not spent on the Indians, except from a kind of half-hearted formality, but reserved for the physical and mercantile appurtenances of establishment:

> Corn-fields, Orchards, Streets Inhabited, and a place of Merchandise cannot denominate *New-England* . . . NEW-ENGLAND is not to be found in NEW-ENGLAND, nor BOSTON in BOSTON.[85]

Scottow's historical sensibility moves at last to the point at which the impossibility of arresting the onward rush of supersessive time becomes acknowledged. The call for repentance and return is disturbed by a countertheme of continual mutation. Without the *magisterium* of a proper presbyterian government (and the lesson of the history is that no such government will stand), writes Scottow,

> *neither the Purity, Order, nor the Beauty, or Glory of the Churches of Christ, nor his Majesty, or Authority in the Government of them, can be long preserved . . . the Churches of old, and late, have Degenerated into Anarchy, or Confusion, or else given themselves up, unto the dominion of some prelatical Teachers to rule at pleasure, which was the poison and bane of the other Primitive Churches; and they will do the same for the future, in the neglect of this Order.*[86]

The history of the church in New England is here explicitly a repetition of the history of the primitive Church; New England reenacts the decline into apostasy, becomes its own antichrist. Here is the Reformation intellect separated not only from the past and the larger world, but also from itself.

Cotton Mather's *Magnalia Christi Americana* expresses, from the perspective of the third generation rather than the first, the same retrospective and nostalgic view of New England's history as does Scottow, but in a far more sophisticated literary and structural form. It begins with a rush of language, a dynamic meter measuring out dynamic history:

> I WRITE the *Wonders* of the CHRISTIAN RELIGION, flying from the depravations of *Europe*, to the *American Strand:* and, assisted by the Holy Author of that *Religion*, I do, with all conscience of *Truth*, required therein by

Him, who is the *Truth* itself, report the *wonderful displays* of His infinite
Power, Wisdom, Goodness, and Faithfulness, wherewith His Divine Providence hath *irradiated* an *Indian Wilderness*.[87]

The language flies with the violent outflung movement to the American strand. The onrush of rhythm resembles Johnson's, but the gram-
matical subject has changed: the mover is no longer divine, neither
God nor devil, as in Johnson and Bradford respectively, but "I." The
seeming confidence of the diction cannot be separated from the act of
remembrance, of "I write." This is a consciousness one step removed
from the history and providences, the *magnalia*, it records, a self-
observing mind for whom history means, at least in part, the literary
form it is inventing, rather than the past itself. All interpretation that
is conscious of itself as willful act has lost something of power and will.
Compared with the works of Bradford and Johnson, the *Magnalia*
declares itself from its first two words to be a secondary epic; the "I
write" strips itself of the objectivism of Bradford and Johnson's form,
which might be expressed as "God does," and bears the same relation
to them as Virgil's "I sing" does to Homer's "Sing goddess."[88]

The "American Strand" inevitably carries the connotation of being
stranded, and the General Introduction is filled with intimations of
exile, of being cast out:

. . . for the worship of God, in a wilderness, in the ends of the earth.

It is the History of these PROTESTANTS, that is here attempted: PROTESTANTS
that highly honoured and affected the *Church* of ENGLAND, and humbly
petition to be a *part* of it: but by the mistake of a few powerful *brethren*,
driven to seek a place for the exercise of the *Protestant Religion*, according
to the light of their consciences, in the desarts of *America*.

It may be, 'tis not possible for me to do a greater service unto the Churches
on the *best Island* of the universe, than to give a distinct relation of those
great examples which have been occurring among Churches of *exiles*, that
were driven out of that *Island*, into an horrible *wilderness*, meerly for their
being well-willers unto the *Reformation*.[89]

"The ends of the earth," "driven" to "the desarts of *America*,"
"Churches of *exiles* . . . driven out of that *Island*, into an horrible *wilderness*"—this is a peculiar accompaniment to Mather's vaunting of
the American church. There is in all of these phrases the consciousness

of a castaway, who hates his forced habitation, who longs to return to that "*best Island* of the universe." It is an example of national religion, of the elect nation, but that nation is England.

If Mather displays a sense of geographical estrangement, his temporal estrangement is even more pronounced. He proclaims a Protestant and also Renaissance, Petrarchan originalism: "In short, the *first Age* was the *golden Age:* to return unto *that*, will make a man a *Protestant*, and I may add, a *Puritan*."[90] At the same time, his idea of history involves a continuation of supersession, of going beyond:

> They zealously decreed the *policy* of complying always with the ignorance and vanity of the *People;* and cried out earnestly for purer Administrations in the house of God, and more *conformity* to the *Law of Christ*, and *primitive Christianity:* while others would not hear of going any further than the *first Essay* of *Reformation*. 'Tis very certain, that the *first Reformers* never intended, that what they did, should be the *absolute boundary* of *Reformation*, so that it should be a sin to proceed any further; as, by their own going beyond *Wicklift*, and changing and growing in their own *Models* also, and the confessions of Cranmer, with the *Scripta Anglicana* of *Bucer*, and a thousand other things, was abundantly demonstrated.[91]

The structure is one of a constant going beyond, which is also a turning back to the first purity of the gospel. The form of Church government is immutable, an atemporal form that cannot be altered.[92] There are, therefore, two kinds of supersession: one is a supersession of previous and imperfect reformation toward a more perfect conformity with primitive perfection, the other a supersession into apostasy. The first kind of supersession implies the possibility of the second. America, writes Mather, was a great secret kept hidden in time until its revelation in the fullness of time. Indeed, God hid three such secrets:

> . . . *three* most memorable things . . . did near the same time, namely at the conclusion of the *fifteenth*, and at the beginning of the *sixteenth century*, arise unto the world: the first was the *resurrection of literature;* the second was the opening of *America;* the third was the *Reformation of Religion*.[93]

The Church, he continues, "must no longer be wrapped up in *Strabo's* cloak."

All of this amounts to a theory of the obsolescence of old forms of knowledge, and their supersession by the new. Mather's historical and epistemological quandary is one of maintaining essentialism within a supersessive model of time. Once the principle of the possibility of a secret future has been set forth, the world is filled with the possibility of further change into inconceivable configurations. This is the second kind of supersession, and the *Magnalia* is haunted by it:

> *Mankind will pardon* me, *a native of that country, if smitten with a just fear of incroaching and ill-bodied* degeneracies, *I shall use my modest endeavours to prevent the* loss *of a country, so signalized for the* profession *of the purest Religion, and for the* protection *of God upon it, in that holy profession. I shall count my country* lost, *in the* loss *of the primitive* principles, *and the primitive* practices, *upon which it was at first established: but certainly one good way to save that* loss, *would be to do something that the memory of the* great things done for us by our God *may not be* lost.[94]

Mather is possessed by this degeneracy, and sees in the history of New England just such a decline. He speculates in the general introduction that Christ may have led the Puritans into the "*dark regions of America*," also described as "*outer darkness*" and "*Tenebrae Exteriores*," only so that they may be a light in the darkness for a brief time, to be an example to the churches of England and Europe, "*a specimen*," "and *this* being done, he knows not whether there be not *all done*, that *New-England* was planted for; and whether the Plantation may not, soon after this, *come to nothing*."[95] He begins the first book with an account of the Huguenot colony in Brazil, the first planting of the true gospel in the New World, but "*unhappy* controversies *arose among them*," and of the fate of those originally pure churches "*no other can be learned, but that they are entirely* lost, *either in paganism or disaster*."[96]

This then is the anticipated fate of New England. More fully than Scottow, Mather feels the full poignancy of the supersessive model of time. He remains an essentialist, committed to the primitive principles and practices, but does not believe that these will persist historically.[97] And derived from this historical perception is his literary self-consciousness. From the tension between his perception of the flux of time and his essentialist desire to stop time comes a literary jeremiad in which the possibility of renewal exists not in history, but only in the text. He will "*prevent the* loss *of a country*," but that prevention consists not in actual renewal, but in writing, "*that the memory of the* great things done for us by our God *may not be* lost." Again, "But whether *New England* may *live* any where else or no, it must *live* in our *History!*"[98] This is an awareness that the literary form of history does not cohere with the structure of time itself, that it does not describe immediate truth; it is interpretation aware of its own interpretive act, and thus exiled from the truth to which it aspires.

The literary style of the *Magnalia* is one of endless digression and concentrated allusion. The allusion is an attempt to place New England, and the *Magnalia* itself, in global context, in a relation to the past, but it only succeeds in a relation of one literary text to other literary texts. There is no sense of any real medium of communion linking age to age or text to text. Indeed, as with Strabo's cloak, Mather often cites in order to reject.[99] This is the opposite of *auctourism*: it is a constant, even compulsive commentary upon the texts of the past, coupled with an intellectual separation from the context of the past. It is the literary structure of the nostalgist, of the antiquarian, and not of the reliquarian imagination.

The density of antiquarian citations serves also as the main subject and means of digression. This digression occurs on almost every page of the *Magnalia* and cannot really be demonstrated here without quotations of inordinate length. The major function and effect of the digressive style, as in Sterne or De Quincey, is to slow down time, to extend the act of telling by introducing an infinity of other material as an exercise of the mind's ability to control and subjectivize the experience of duration.

A similar motive is evident in the larger structure of the *Magnalia*. The first book opens with the Brazilian colony, then turns backward to the early discoveries of America, then to the founding of the Plymouth colony (these events Mather calls "Primordia"), from which point it gives a brief history of the whole province up to 1696, ending with a list of the ministers and churches in that year, an "*Ecclesiastical Map* of the country." The first book thus ends with a static inventory, but

> Know then, that ... for more than twenty years, the *blasting strokes* of Heaven upon the secular affairs of this country have been such, as rather to abate than *enlarge* the growth of it.[100]

This is followed by a long digression, actually a reprinting of one of Mather's lectures, on the history of Boston, containing little history, but much effusion on God's grace to the city. The second book opens with a life of William Bradford, goes on to his successors, and then turns back to a life of John Winthrop, "Nehemias Americanus," gives an account of his successors, then turns back again to a life of Simon Bradstreet, "Pater Patriae." This pattern continues through the first three books, and is then followed by a history of Harvard, then more lives of "some eminent persons educated in it." The fifth book contains a history of the synods of the New England church and their doctrinal pronouncements. Book six is an anecdotal record of divine providences, and book seven a history of the wars with the Indians, "A Book of the Wars of the Lord." This structure of dividing the history into discrete subjects, particularly into lives, which allow almost infinite subdivision, gives to the *Magnalia* a quality of endlessly turning backward. After the life of one hero of the primitive foundation, linked to the present by his successors, Mather can turn back once again to the beginning, to give another biography, or a history of the founding of Harvard. The whole form is a kind of digression, an effort to stop time, to return to that primitive point of the foundation, before the degeneracies, before the present possibility that it will all "come to nothing."

This is the conflict throughout the *Magnalia*: between on the one hand a sense of the violent supersessive onrush of time, defined as mutation and exile, which will not allow any certainty, which reduces all essentialist attempts to establish eternal truth to degradation and dissension, which supersedes all the superseders, and on the other hand the literary effort to fix time in a series of memory pieces, timeless exemplary biographies. The form is the antithesis of Bradford's annals, which surrender history to time. Mather has a more tenacious hold upon earth than Bradford, but his very tenacity estranges him from the present and the choric;[101] Bradford retreated from the world into the divine, Mather from the world into the text.

In the last book of the *Magnalia* the literary form of reversion breaks down in a story of declension, concluding in a ritual jeremiad. At last Mather's strategies of delay bring him to his own present and to an uncertain future. Or, rather, the literary strategy becomes identified with the jeremiad as its social consensual equivalent. The very last sentence of the book makes the connection explicit: "*Let us search and try our ways, and turn again unto the Lord.*"[102] But the ritualized turning-again negates the very idea of religious establishment (this is a problem inherent in all of the uses of the jeremiadic form in Puritan rhetoric); it is an acceptance of the inevitability of decline from the first love, and of the need for new reformation, for which Mather calls in the last pages of his history. The very institutionalization of the jeremiad is a concession to and recognition of the problem of mutation, and the very call for reform emphasizes that mutation.

By the time of Jonathan Edward's *History of the Work of Redemption* (published 1773), the turning-again of the jeremiad had become the repeated charismatic reformations of revivalism, no longer a societal imperative, but a series of mysterious pourings-out of the Spirit.[103] The Pentecostal fire is a fitful blaze in a matrix of darkness, a Reformed Church that

> is much diminished . . . altered for the worse from what was in the former times of the Reformation [in] the prevailing of licentiousness in principles and opinions. . . . Arianism, and Socinianism, and Arminianism, and Deism, are the things which prevail, and carry almost all before them. And particularly history gives no account of any age wherein there was so great an apostacy of those who had been brought up under the light of the gospel, to infidelity; never was there such a casting off of the Christian and all revealed religion; never any age wherein was so much scoffing at and ridiculing the gospel of Christ, by those who have been brought up under gospel-light, nor any thing like it, as there is at this day.

> But now there is an exceeding great decay of vital piety; yea, it seems to be despised, called *enthusiasm, whimsy,* and *fanaticism.* Those who are truly religious, are commonly looked upon to be crack-brained, and beside their right mind; and vice and profaneness dreadfully prevail, like a flood which threatens to bear all down before it.[104]

The two passages reveal a very different sensibility from Bradford's horror and pity at the trial of Thomas Granger, or from Johnson's, Scottow's, and Mather's injunctions to intolerance and persecution of heretics. These all inveigh against the intrusions of (perceived) evil, whether from buggers or from Quakers, into a Church society, threatening to destroy its purity. The essential sin attacked by the jeremiad is the toleration of evil, and the community is exhorted to return to the high ground of intolerance. In Edward's case, the perspective is that of an embattled Christian within a secular state.[105] The work of God continues, but in corners, and in moments. Far from calling for intolerance, Edwards makes a careful tally of the ways in which the world has altered for the worse and for the better since the Reformation, and lists among the better that "There is far less persecution now than there was in the first times of the Reformation."[106] An orthodox remnant of what was once New England's true faith and foundation, he is in no position to counsel intolerance, for he is more likely to be its victim at the hands of an unorthodox majority, than to be its user in the enforcement of conformity to truth.

Of all the elements contributing to the diminution of the Reformed Church, the worst for Edwards is Deism, which he recognizes as being different in kind from other heresies:

> The Deists wholly cast off the Christian religion, and are professed infidels. They are not like the Heretics, Arians, Socinians, and others, who own the scriptures to be the word of God, and hold the Christian religion to be the true religion, but only deny these and these fundamental doctrines of the Christian religion: they deny the whole Christian religion. Indeed they own the being of God; but deny that Christ was the son of God, and say he was a mere cheat; and so they say all the prophets and apostles were; and they deny the whole scripture. They deny that any of it is the word of God. They deny any revealed religion, or any word of God at all; and say, that God has given mankind no other light to walk by but their own reason. These sentiments and opinions our nation, which is the principal nation of the Reformation, is very much over-run with, and they prevail more and more.[107]

"They prevail more and more"—one thinks of Benjamin Franklin at George Whitefield's sermon, walking backward until he reaches the limit of his hearing, and then calculating the greatest number who can be addressed by the human voice. For Franklin this was the only interest of the sermon; he reports nothing of what Whitefield said, and apparently did not tremble for his own soul. Only the physical properties of the voice interested him. It is the perfect image of Deism and Protestantism in tolerant and uncomprehending proximity, of "multitudinarism."[108]

When Protestant historiography reversed at a stroke the meaning of the Christian past, when it declared that secrets had been hidden in time, that all had lived in ignorance from the fifth century to the fifteenth, it created a great schism in the fabric of the perceived world.

For the first time since the formulation of the Christian understanding of time by Eusebius, Augustine, and Orosius, a new thing had been discovered in a world of knowledge that depended for its existence on the proposition that everything that could be known was known. The Reformation sense of time initiated a principle of revolution and of supersession, and made time dynamic. After the great schism with the past had been described and justified, and history reinterpreted in its light, the task of the subsequent generations of Reformed historians was to reestablish stasis, for without a stasis of knowledge no essential truth could endure. Of this dilemma and task, the Puritan historians of New England provide an example in which the forms can be clearly seen, extracted from the many conservative and radical complexities of Europe, concentrated by isolation and simplicity into the focal point of the burning glass. Theirs was a controlled experiment, and in it the future inevitabilities of European history can also, and most clearly, be seen. Cut off from the past and from the non-Protestant world, they waged an ideological warfare of containment: the newly discovered (but original, unchanging) truth is now established; there must be no new revolutions, no new things discovered, no further schisms within the schismatic body, for the greatest fear of the supersessor is to be superseded.

The fear of the reformed church-state was the fear of a prince who has gained his throne by assassination: he has himself seeded his dominion with the precedence for his destruction, and for the destruction of a thousand usurpers after him, each one reigning for a shorter time. Once the delicate pattern of the past had been destroyed, then time began to move in the human consciousness. Those who had moved it would have it stop after a single adjustment, but the new paradigm of history carried its own inexorable logic. It propelled Joshua Scottow, Cotton Mather, and, much more so, Jonathan Edwards (and the differences between them are generational, not doctrinal or temperamental) into a truly new world, neither anticipated nor desired.

Epilogue: Accelerations

Vladimir: That passed the time.
Estragon: It would have passed in any case.
Vladimir: Yes, but not so rapidly.

—SAMUEL BECKETT, *Waiting for Godot*

In 1836 Emerson began the first page of his first book with a complaint about time:

> Our age is retrospective. It builds the sepulchres of the fathers. It writes biographies, histories, and criticism. The foregoing generations beheld God and nature face to face; we, through their eyes. Why should not we also enjoy an original relation to the universe? Why should not we have a poetry and philosophy of insight and not of tradition, and a religion by revelation to us, and not the history of theirs? Embosomed for a season in nature, whose floods of life stream around and through us, and invite us by the powers they supply, to action proportioned to nature, why should we grope among the dry bones of the past, or put the living generation into masquerade out of its faded wardrobe?[1]

Or again, this from the address "The American Scholar," delivered a year after the publication of *Nature*:

> Our day of dependence, our long apprenticeship to the learning of other lands, draws to a close. The millions that around us are rushing into life, cannot always be fed on the sere remains of foreign harvests. . . . Each age, it is found, must write its own books; or rather, each generation for the next

149

succeeding. The books of an older period will not fit this. . . . Meek young men grow up in libraries, believing it their duty to accept the views which Cicero, which Locke, which Bacon, have given, forgetful that Cicero, Lock, and Bacon were only young men in libraries when they wrote these books.[2]

The complaint is that the present is in bondage to tradition, and what is proposed is a revolution of original relation to nature. This much is obvious, but more subtle fish swim beneath the surface of a pond less clear and far deeper than Walden. Emerson's call is not to a new relation to the universe, but to the recovery of an ancient perfection of insight ("The foregoing generations beheld God and nature face to face; we, through their eyes.") from which the present is degraded. The primitive past had a perfect form of worship in that it beheld nature atemporally, through immediate and mystical apprehension; the closer and historical past, which is contiguous with Emerson's own time and society, is encased in a carapace of tradition that prevents immediate perception. This seems to be a criticism of a traditionalist age, but Emerson's age cannot, even in his own perception of it, be traditionalist. The mark of a truly traditionalist culture is that for it the past is an enabling force, which cannot even be perceived as foreign to the present; for Emerson the past is oppressive, a confining and crippling force: "They pin me down. They look backward and not forward." If, indeed, Emerson's society were truly traditionalist, truly bound to the past as he accuses it of being, he would not be capable of perceiving the pastness of the past, its difference from the present that makes its veneration confinement and denial of the present, the generation "rushing into life."

Such a complaint of retrospection can only originate within a culture that senses itself cut off from the past and yet still enslaved to it, a culture whose model of time is one of epistemological supersession, of new forms of knowledge superseding and making irrelevant the old. To progress from a traditionalist paradigm to the complaint and revolution Emerson proposes here requires many intermediate stages; the whole process of the disintegration of the medieval and traditionalist model of time from Petrarch and Luther forms the long foreground to Emerson's revolution, as it does to the Puritan histories of Massachusetts Bay.

Seen in this continuum, the complaint of retrospection conforms to the established model of Reformation history. It makes war on the immediate and historical past in order to unite the present to that fugitive moment of perfection in the mythical and distant past, when mankind had an original relation to the universe. Building the sepulchers of the fathers is not different from the "vile ceremonies of men," "the human inventions" that so incensed the Reformers. The assertion is that all forms of human institution, of cultural tradition, are illicit precisely because they are humanities, and therefore degraded from the direct, transcendent perception of God. The past, however, remains a haunting, potent force in the mind that must be invoked if essential knowledge is to remain a possibility, and revolution cannot occur without the possibility of establishing essential truth. Hence the admixture of devotion to and estrangement from the past, reconciled by separating the past from history, the ideal moment of original unity from the parenthesis of antichrist's time, the dynamic onrush of unknowledge in the realm of the simulacra.

The new revolution is therefore not against a traditionalist culture, but against the previous stage of revolution, declared here to be traditionalist; the "sepulchers of the fathers" inevitably suggest relics, veneration of the saints, cults of the martyrs—all that the Reformation would brand as blasphemous idolatry. Yet the culture so characterized in these very Protestant terms of complaint is already one of degraded Protestantism, greatly removed even from the first stages of revolution. The stages can be seen in Emerson's Unitarian ministry: Unitarianism is a compound of Puritan Congregationalism with rationalistic Deism and Arianism, a combination of elements Cotton Mather could not have imagined in his darkest premonitions of New England's decline. Its direct descent from Puritanism may be illustrated by the fact that the two original Puritan churches in Boston, the First Church, founded in Boston proper (the Shawmut Peninsula) in 1630, and the Second Church, founded in the growing North End in 1649, both became Unitarian, and still exist as the combined First and Second Church in Boston, on Marlborough Street in the Back Bay. Unitarianism represents a rationalist revolution within and against Puritanism; it maintains the structure of worship and church governance while abhorring the content of that structure: biblical, credal, and supernatural belief—all that the eighteenth century reproved as the appurtenances of sectarian religion.[3] These are also, of course, historical appurtenances. Within this church or antichurch in rebellion— and the foundation of the American Unitarian Association in 1825 provides a convenient date at which to place the revolution[4]—Emerson could not reconcile himself to the few vestigial remnants of doctrine, nor to the offensive symbolism (by this time a desacralized mere symbolism) of the sacraments.

Within this context, the foundation of Emerson's transcendentalism can be seen as an act of conformity with the prevailing historical paradigm (the same can be said of the European transcendentalisms of Kant, Fichte, and Hegel). This paradigm calls for continued generational revolution. In the terms established as the model for Protestant autobiography by Bunyan's *Pilgrim's Progress*, the Celestial City of the previous revolution, the goal of its establishment, becomes the City of Destruction for the new, the place that must be renounced in order that a new consciousness can begin: so Rousseau commences by being locked without the gates of Calvinist Geneva; so Franklin, a childhood devotee of *Pilgrim's Progress*, exiles himself from Puritan Boston in order to establish the rational city in Philadelphia, a place of human, not divine, concord. Each new revolution finds itself in rebellion against the historical material retained by the previous revolution, and the historical material decreases at each stage. This form of revolutionary paradigm grows by what it feeds upon: each stage of revolutionary supersession makes the past more sectarian, more schismed, more multitudinous. Each historical attempt to establish a universal, essential truth can be seen as one more sect, one more schism, one more pathetically temporary separatist discourse of words led out to do battle against other words. As Thoreau puts it almost with glee:

> What everybody echoes or in silence passes by as true to-day may turn out to be falsehood to-morrow, mere smoke of opinion, which some had trusted for a cloud that would sprinkle fertilizing rain on their fields.... One generation abandons the enterprises of another like stranded vessels.[5]

Augustinian history answered a condition of epistemological crisis with a unity of historical knowledge. Only as long as the unity of the historical paradigm was maintained could it demand allegiance. Once history became the province of factions, it became invalid as the source of knowledge, a crippling rather than enabling force, a demonic and uncanny haunting resented by those who, by the very conditions of human life and consciousness, cannot escape its presence. For Thoreau, therefore, the by-now-apparent fleetingness and inconsistency of what the past has held true becomes a liberator from the claims of historical knowledge. This can be a delight because he already has an essential knowledge with which to replace the historical: the immediate perception of nature. Yet such a perception actually resides in extreme linguistic assertion, in Emerson's claim that

> Undoubtedly we have no questions to ask which are unanswerable. We must trust the perfection of the creation so far as to believe that whatever curiosity the order of things has awakened in our minds, the order of things can answer.[6]

As with the encyclopedic quest of the Enlightenment, the questions must be answered before they can be asked, the goal of progress and search must be defined in present terms before progress can be contemplated or the search begun, nature confined to Thoreau's "quiet parlor of the fishes, pervaded by a softened light as through a window of ground glass" in Walden pond—that wonderfully domesticating image—and not allowed its infinite expansion into the unhumanness and unknowableness of Melville's ocean.

If Emerson's complaint conforms to the prevailing paradigm of the constant supersession of paradigms, then it also conforms to the paradigm of establishment transmitted by Reformation historiography. It makes parenthetical the historical past, which by now denotes the establishment of previous revolutions, and unites ahistorical present with remote ahistorical past in an establishment of new essentials.[7] Nature represents, as did the Bible and reason before it, an escape from the difficulty of historical knowledge into a realm of knowledge not susceptible to the forces of historical mutation and supersession. But the assertion of that knowledge was the product of, was in fact *made of*, a historical and human language. Transcendentalism, like all attempts to establish the essential since the sack of the great city of Augustinian history, was a temporary dialect.

Emerson represents a subliminal transmission of and participation in the paradigm. Even Thoreau's "one generation abandons the enterprises of another like stranded vessels"—closer to full realization—is mediated by the perceived exception of transcendentalism from the overall movement. Full consciousness of the structure of dynamic and supersessive history, and of its effect on epistemology, comes when the latest revolution is no longer believed in, when the faith that here at last is an essential establishment that will stop the flux and endure cannot be maintained. An illustration of this sensibility can be found in Arthur Hugh Clough, a correspondent of Emerson and an admirer, but never sure enough to be a disciple of any movement:

> Ah my friends[,] gravitation is discovered, and behold—law within the law, a something that is interior to it; that comprehends, and other things,—an attraction of attractions,—who can say?—begins to be talked of! Ptolemy in old times thought he had made it out, and Ptolemaic theories perish with the long ages of puzzle; and victorious over cycle and epicycle, behold, the perfect Newtonian. Which explains all!— And the world has not done congratulating itself on moving in an ascertained ellipse around an established centre of all, when lo, the centre is no centre, there is another somewhere; a centre of centres; it is not the sun now, but in the constellation Hercules or something or other. We touch the line which we thought our horizon; it was [a] line of shadow which we enter and discern not. We approach, and behold, leagues away, and receding and receding yet again beyond each new limit of the known a new visible unknown. You have found out God[,] have you? Ah my friends! let us be—*Silent*.[8]

Before such a vision of the past there are only two choices: either a renewed assertion of the essential, which can deny the past under a rubric of progress or evolution, or a surrender to supersession, a realization of the temporality of all attempts to describe the world, that neither the present nor the future synthesis will hold. The first is an arrogance, exemplified by Gibbon on the gothic cathedral: "I darted a contemptuous look on the stately monuments of superstition."[9] The second will allow no surety, no belief in the truth of any paradigm other than the structure of change itself. It is the end of all statement, all assertion: "Ah my friends! let us be—*Silent*."

Another skeptical half-disciple of Emerson, Nietzsche, writes of

> der *historischen Krankheit*. Das Uebermaass von Historie hat die plastische Kraft des Lebens angegriffen, es versteht nicht mehr, sich der Vergangenheit wie einer kräftigen Nahrung zu bedienen.

> (the *historical malady*. The excess of history has attacked the plastic powers of life, it no longer understands how to avail itself of the past as heart nourishment.)[10]

What Nietzsche means by history is scientific history, that is, the history of process and supersession, which he also takes to be objective history:

> ... *durch die Wissenschaft, durch die Forderung, dass die Historie Wissenschaft sein soll* ... regiert nicht mehr allein das Leben und bändigt das Wissen um die Vergangenheit: sondern alle Grenzpfähle sind umgerissen und alles was einmal war, stürzt auf den Menschen zu. So weit zurück es ein Werden gab, soweit zurück, ins Unendliche hinein sind auch alle Perspektiven verschoben. Ein solches unüberschaubares Schauspiel sah noch kein Geschlecht, wie es jetzt die Wissenschaft des universalen Werdens, die Historie, zeigt: freilich aber zeigt sie es mit der gefährlichen Kühnheit ihres Wahlspruches: fiat veritas pereat vita.

> (... *through science, through the demand that history be a science* ... life is no longer the sole ruler and master of knowledge of the past: rather all boundary markers are overthrown and everything which once was rushes in upon man. All perspectives have shifted as far back as the origins of change, back into infinity. A boundless spectacle such as history, the science of universal becoming, now displays, no generation has ever seen; of course, she displays it with the dangerous boldness of her motto: *fiat veritas pereat vita.*)[11]

"Let there be truth, and may life perish"—the history of change and difference destroys the possibility of life in the present, destroys the possibility of any assertion, because it makes all culture relative and passing:

> Der historische Sinn, wenn er *ungebändigt* waltet und alle seine Consequenzen zieht, entwurzelt die Zukunft, weil er die Illusionen zerstört und den bestehenden Dingen ihre Atmosphäre nimmt, in der sie allein leben können. ... Eine Religion zum Beispiel, die in historisches Wissen, unter dem Walten der reinen Gerechtigkeit, umgesetzt werden soll, eine Religion, die durch und durch wissenschaftlich erkannt werden soll, ist am Ende dieses Weges zugleich vernichtet. Der Grund liegt darin, dass bei der historischen Nachrechnung jedesmal so viel Falsches, Rohes, Unmenschliches, Absurdes, Gewaltsames zu Tage tritt, dass die pietätvolle Illusions-Stimmung, in der Alles, was leben will, allein leben kann, notwendig zerstiebt: nur in Liebe aber, nur umschattet von der Illusion der Liebe schafft der Mensch, nämlich nur im unbedingten Glauben an das Vollkommene und Rechte. Jedem, den man zwingt, nicht mehr unbedingt zu lieben, hat man die Wurzeln seiner Kraft abgeschnitten: er muss verdorren, nämlich unehr-

lich werden. In solchen Wirkungen ist der Historie die Kunst entgegengesetzt: und nur wenn die Historie es erträgt, zum Kunstwerk umgebildet, also reines Kunstgebild zu werden, kann sie vielleicht Instincte erhalten oder sogar wecken. Eine solche Geschichtschreibung würde aber durchaus dem analytischen und unkünstlerischen Zuge unserer Zeit widersprechen, ja von ihr als Fälschung empfunden werden.

(The historical sense, if it rules *without restraint* and unfolds all its implications, uproots the future because it destroys illusions and robs existing things of their atmosphere in which alone they can live. . . . A religion for example which, under the rule of pure justice, is to be transformed into historical knowledge, a religion which is to be thoroughly known in a scientific way, will at the end of this path also be annihilated. The reason is that the historical audit always brings to light so much that is false, crude, inhuman, absurd, violent, that the attitude of pious illusion, in which alone all that wants to live can live, is necessarily dispelled: only with love, however, only surrounded by the shadow of the illusion of love, can man create, that is, only with an unconditional faith in something perfect and righteous. Each man who is forced no longer to love unconditionally has had the root of his strength cut off: he must wither, that is, become dishonest. In such effects art is opposed to history: and only if history can bear being transformed into a work of art, that is, to become a pure art form, may it perhaps preserve instincts or even rouse them. But such a manner of writing history would thoroughly contradict the analytic and inartistic trend of our time, it would even be perceived as falsification.)[12]

Nietzsche does not question the objectivity or factuality of scientific history; he protests that it is a truth contrary to life because it deprives mankind of belief, reduces it to Clough's silence. Nietzsche advocates instead a return to art or love, or the conditions of the unhistorical and the superhistorical—the first a deliberate enclosing of oneself within a horizon, a deliberate ignorance of history, the second a perception of being in a Platonic or Christian sense, a philosophical realism.[13] He rejects attempts to shape the process, because these are fraudulent accommodations; they can only accept the process by confining its apex to their own image, thus

Dieser Gott aber wurde sich selbst innerhalb der Hegelischen Hirnschalen durchsichtig und verständlich und ist bereits alle dialektisch möglichen Stufen seines Werdens, bis zu jener Selbstoffenbarung, emporgestiegen: so dass für Hegel der Höhepunkt und der Endpunkt des Weltprozesses in seiner eigenen Berliner Existenz zusammenfielen.

(this God [the immanent process-God of Hegelian history] became trans-
parent and intelligible to himself inside the Hegelian craniums and has
already ascended all possible dialectical steps of his becoming up to that
self-revelation: so that for Hegel the apex and terminus of world history
coincided in his own Berlin existence.)[14]

The same accusation applies to Marx, to Macaulay, for whom history culminated in Victorian England, to Bancroft, for whom rational
and democratic America represented the apex and terminus. For the
historians of process, the dialectic stops with their own culture (except
in the case of Marx, whose terminus is future, but this future is close
and defined, and beyond it the social dialectic must not mutate). This
defined culture then becomes the *novus ordo seclorum*, the new order
of the ages for all cultures. The histories of progress not only assert the
present over the past, but also their particular national and, by extension, racial culture over all others. This is another paradigm transmitted from Reformation historiography, and one I have explicated in
most detail in my treatment of Edward Johnson's "burning-Glasse."
The culture-in-isolation universalizes itself through a model of ideological conquest, which conquest has been preordained by God for the
salvation of the world. Such salvation inevitably implies a contempt
for those to be saved; it can be seen in the Puritan attitudes toward the
Indians and can be found in Bancroft unchanged, except that the criteria for salvation are now cultural rather than Christian:

> Before that time the whole territory was an unproductive waste. Through-
> out its wide extent the arts had not created a monument. Its only inhabi-
> tants were a few scattered tribes of feeble barbarians, destitute of commerce
> and of political connection.[15]

The passage illustrates the inevitability of the connection of race and
progress. To the Middle Ages the wild men were merely elsewhere and
other, intriguing boundaries for the *mappa mundi*. Even in the seventeenth and early eighteen centuries, the Puritans saw the natives of
their hated wilderness as other, children of Satan hidden away from
the gospel, but not as primitive, archaic. Only when history could be
seen as a process of change and supersession could existing peoples be
seen as a prior stage in a process, as obsolete (Vico was perhaps the
first complete articulator of a theory of social development from a savage state, although the idea of savagery as a temporal condition is

found in Hobbes and Locke).[16] We have forgotten the subtitle of Darwin's *Origin of Species: On the Origin of Species by Means of Natural Selection, Or the Preservation of Favoured Races in the Struggle for Life*.[17] Once the dynamic model of history dominates, then cultural difference inevitably becomes temporal difference. Both feed upon each other; the consciousness of difference or separation from the past invokes the consciousness of difference from other cultures. When such a temporally and culturally schismatic mode of thought predominates, the inevitable result is an attempted reconstitution of the lost past as process, development; like geographical colonization, such a temporal colonization is both a possession and a destruction. When the two modes of schism become fully fused, the developmental histories of Gobineau and Chamberlain,[18] and their notorious effects upon the modern world, result.[19]

But if the assertion of a present ideology as the terminus and apex of a historical process is a fraudulent colonization of the past, the only alternative is Clough's silence, a surrender to mere process that is a cultural paralysis. Because of the burden of its supersessive history, writes Nietzsche, modern culture is no culture, but only a knowledge about cultures.[20] Nietzsche's opposition to such a history is necessarily futile. He can revolve only within his opponent's orbit because his mind can only conceive of the evils of supersessive history within the context of supersessive history. History means for him a "science of universal becoming"; he has no sense that history can be anything other, unless it should be a pure art, by which he means a creative abandonment of truth. He has no doubt that what he designates history-as-science possesses the truth, a despairing and soul-destroying truth, and that love and art, which he opposes to the absolute relativism of the science of universal becoming, are illusions: history

> destroys illusions and robs existing things of their atmosphere in which alone they can live . . . the attitude of pious illusion, in which alone all that wants to live can live, is necessarily dispelled . . . only with love, however, only surrounded by the shadow of the illusion of love, can man create, that is, only with an unconditional faith in something perfect and righteous.

The contradiction is heartbreaking. The argument surrenders all truth to the enemy from the outset, and yet still condemns the truth. Nietzsche asserts the need for an active and humane culture to do away with history by means of a Platonic lie-in-need.[21] But this is another form of essential assertion, even further removed from the validity of historical knowledge. It is the assertion of a truth that knows itself to be a lie, the burden of Dostoevski's Grand Inquisitor, and indicative of a horrifying degree of cultural displacement. Love and art can only exist as self-aware solipsisms, asserted against all known truth, and culture—and, once again, Nietzsche is trapped within the paradigm he hates by emphasizing a national, German culture—as a communal solipsism.

I offer Emerson, Thoreau, Clough, and Nietzsche as examples of the persistence and power of the Reformation historical paradigms—displayed in various degrees of subconscious, semiconscious, and conscious transmission—and of the various possible responses to the dynamic perception of time. My excuse for grouping them in this fashion is the well-established and undeniable chain of influence between the four. It says much for the perceived cultural incoherence of an age that such an excuse for connecting Massachusetts to England and Germany should be necessary.

For both Clough and Nietzsche, the driving force of dynamic supersession is science: for Clough the constantly changing image of the world wrought by natural science, for Nietzsche the scientific knowledge of the past which destroys all "illusion." Vico makes a distinction between science (*scienza*) and consciousness (*coscienza*).[22] By the former he means a knowledge of the true (*il vero*), derived from philosophy; by the latter a knowledge of the certain (*il certo*), of the shared knowledge of human institutions, arts, languages, societies—everything that might be grouped under the humanities—knowledge of these being derived from philology, the study of languages and texts. But the new science of Vico's title comprehends both of these; it aspires to a knowledge of what is universal and eternal (*il vero*) in the development of human society; "science," Vico quotes Aristotle, "has to do with what is universal and eternal (scientia debet esse de universalibus et aeternis)."[23] What is proposed is a science of dynamic history: a world of mutation and supersession whose principles are nevertheless unchanging. The effect is a complex attempt to redeem history from the burden of historical knowledge. Vico radically reinterprets his textual sources in accordance with the principles of his new science. The work is both a denial of the past in the terms of that past's self-conscious knowledge, embodied in the texts it has transmitted to

the present, and an attempt to redeem the chaos of a past now perceived to be a record of mutation, by eternal and transcendent principles. It is no less than a scientific history, an attempt to escape historical history. Although Vico's work is in many ways opposed to
rationalism and to the Enlightenment (Montesquieu's *The Spirit of
Laws*[24] is the first scientific history fully in accord with the Enlightenment), it nevertheless fits (indeed, founds) Cassirer's characterization
of Enlightenment history:

> Die Geschichte ist es, die hier der Aufklärung die Fackel vorträgt; die die
> "Neologen" aus den Banden der dogmatisch gedeuteten Schrift und der
> Orthodoxie der vorangehenden Jahrhunderte erlöst.

> (History bears the torch for the Enlightenment; it frees the "neologist" from
> the bonds of Scripture dogmatically interpreted and of the orthodoxy of the
> preceding centuries.)[25]

The application of the term "science" to history, and indeed to the
physical world, is a reversal of the word's original meaning, as a glance
at the *Oxford English Dictionary* readily indicates. "Science" desig-
nated true knowledge, derived from philosophy, particularly from
metaphysics; its subject was God, the Good, Being. It was particularly
used to designate God's own knowledge. The transference of "science"
to the physical world of flux, the Platonic opaque-to-reason, elevates
the empirical to the absolute, and consigns theology and metaphysics
to the world of mere paradigms. One of the early conjunctions of "science" with "natural" occurs in Hobbes, and the shift in the ground of
knowledge, and the motivation for the shift, can be seen clearly:

> ... in this natural kingdom of God, there is no other way to know any
> thing, but by natural reason, that is, from the principles of natural science;
> which are so far from teaching us any thing of God's nature. . . . And there-
> fore, when men out of the principles of natural reason, dispute of the attri-
> butes of God, they but dishonour him.[26]

This is the rationalist disgust with sectarian religion, an ur-Deistic
wish to say as little about God as possible, as a means of social survival. It is not different from the motivation of Calef and Brattle in
Massachusetts. Religion had become paradigmatic by its multiplicity
and contentiousness, both within the synchronous social world of reli-

gious wars and persecutions, and within the diachronous world of supersessive history, of the strife of one age with another. Casting aside all speculation about God, Hobbes would find concord in the world of appearances, where the communality of sensation can provide ground for agreement.

This is the foundation of the modern understanding of "science," of a true knowledge that can be found only in physical things. Transferred to this new realm, "science" still carries its former meaning of true knowledge, fact, not a paradigm. It is understood in this sense both by Clough and Nietzsche, for whom natural science and scientific history have become the drivers of the sense of accelerating supersession that leads both to epistemological and cultural despair. The supersessive shifting of descriptions of the world—the scientific history that describes the principles and shapes of constant supersession—leaves no possibility for a resting place for the mind. Only by an assertion that knowledge is essentially complete—the Hegelian or Macaulayan or Bancroftian idea of progress, which defines the present as the *telos*—can the supersessive be transmuted into the pattern of progress. Neither Clough nor Nietzsche can accept such a facile limit to the history of supersession. That this history itself is paradigmatic does not occur to them; such is the inherited incantatory power of the word "science" that Nietzsche can find solace only in the maddening self-conscious assertion of *illusions* of love and truth and coinherence of past with present. The nineteenth century was greatly concerned with "natural science" and so invested it with "science" and saw it as the driving force of historical supersession; the fourteenth to seventeenth centuries were primarily concerned with religious questions, invested them with "science," and saw God as the driving force of historical supersession. As I have demonstrated, in the formation of the modern dynamic and supersessive sense of historical time, the religious motives have the priority, being far more powerful in their convulsion of the world of knowledge than Copernicus's and Galileo's cosmology, or Tycho Brahe's nova. The rise of science is not the cause of the new model of history, but is caused by it, and is an attempt to reach accommodation with it. All areas of knowledge that become identified with the supersessive pattern tend to be reduced to mere paradigms, and it is this reduction that the term "science" is designed to work against. Religion in the view of the Enlightenment rationalists was so reduced; science, in the work of Kuhn, is undergoing a similar reduction.[27]

In New England, Thomas Prince's *Chronological History of New England* (Boston, 1736) marks the turning from religious to "scientific" history. As the title indicates, it is a work concerned with the new chronologies of Scaliger and Ussher [28] (which chronologies, although they represented a new rationalization and reconciliation of the chronological tradition, as does the chronology that commences Vico's *Scienza nuova*, according to the laws of supersessive history were consigned to ludicrous antiquity by the chronological revolution of the nineteenth century), and tells the history of the world from the sixth day of Creation to 1633. Prince values most "scientific" accuracy of dating and historical fact above any rhetorical purpose or any reading of divine providence, and his history abdicates the responsibility of creating historical consciousness. Also, despite its title, New England is the suppressed material of the work; it cannot comprehend the history of the colony after that moment of establishment, and concludes its account in 1633. Breisach's comment is apposite:

> Unfortunately, the work's merits were defeated by its tedium. It did signal, however, a radical change in perspective. The proclamation of the New Zion had been replaced by a narrative which confirmed Newton's mechanistic universe.[29]

Prince represents the exhaustion of the religious rhetoric of history, this exhaustion being expressed in a history of fact without ideological or teleological purpose. Rhetorical, purposive history was recovered in the revolution, in the histories of David Ramsey, Jedidiah Morse, and Noah Webster, in Mason Locke Weems's biography of Washington,[30] and culminated in Bancroft. The purpose was to fix America as the *telos* of historical development and the salvation of the nations, the *novus ordo seclorum*. This purpose is of course entirely part of the structure of Reformation history as I have described it. It is the assertion of national history as the universal, of the present as the end of progress. It is the self-glorifying and conservative history of revolutionaries afraid of continued revolution, as in Bancroft's almost medieval originalism:

> Other governments are convulsed by the innovations and reforms of neighboring states; our constitution, fixed in the affections of the people . . . neutralizes the influence of foreign principles.[31]

A historical system decays when it can no longer contain the phenomena of its culture, and the idealist, progressivist, one might call them Comtist, histories of the revolution could contain neither the Civil War nor the Gilded Age, despite Bancroft's assertion to the contrary in the note he added to his introduction in 1882.[32] The next movement of American history was that of the critical scientific history associated with the foundation of the American Historical Association in 1884, and with the founding of the *American Historical Review* in 1895. This bears to Bancroft the relation that Prince bears to Johnson or Mather; it represents the exhaustion of a rhetorical mode reduced to mere paradigm, but without a new revolution yet in full sight. Scientific history in this sense abandons the purposive formation of historical consciousness, and retreats into the world of "science," of facts and monographs.

Henry Adams represents a late defection from this school. The mode of defection is a complex one, and I will conclude with *Mont Saint Michel and Chartres* and *The Education* considered, as they should be, as a single compound work of the philosophy of history, a diptych of two tremendously dissimilar panels, one an art of space, the other an art of time. Together they express perhaps better than any other work what it is like to live within the supersessive model of history, at the commencement of the twentieth century.

Consideration must begin with the scheme of the work:

> Any schoolboy could see that man as a force must be measured by motion, from a fixed point. Psychology helped here by suggesting a unit,—the point of history when man held the highest idea of himself as a unit in a unified universe. Eight or ten years of study had led Adams to think he might use the century 1150–1250, expressed in Amiens Cathedral and the works of Thomas Aquinas, as the unit from which he might measure motion down to his own time, without assuming anything as true or untrue, except relation. The movement might be studied at once in philosophy and mechanics. Setting himself to the task, he began a volume which he mentally knew as "Mont Saint Michel and Chartres: a study of thirteenth-century unity." From that point he proposed to fix a position for himself, which he could label: "The Education of Henry Adams: a study of twentieth-century multiplicity." With the help of these two points of relation, he hoped to project his lines forward and backward indefinitely.[33]

Mont Saint Michel and Chartres, the left panel of the diptych, is controlled by spatial, not by temporal forms. It is a tour of the monuments of medieval France, addressed by an uncle to a putative niece, from which the narrative of travel is missing. The niece never speaks;

there are no mechanisms of movement, no engagement of cabs or hotels. These are not Stoddard's *Lectures,* nor *The Innocents Abroad.* The movement is like that of the gliding camera, seemingly suspended without limitation. It can circle the archangel standing on the summit of the tower of Mont Saint Michel, hundreds of feet above the sea, explore the cloister, or glide across France to Amiens or Notre-Dame de Paris. Such a movement is freed from the temporal, because the journey can be redoubled, repeated, reversed at any time without effort, like a journey recalled in dream or memory, but not recalled, actually undertaken.

The very first sentence is spatial—"The Archangel loved heights."[34]—and it contains the two most important spatial forms of the book: the vertical aspiration to height, and the arch, which reconciles the horizontal and vertical in delicate balance. Indeed, the Archangel Michael, with whose figure the book begins, is for Adams primarily the angel of the arch.

The book is to be a study of the medieval world in which "one knew life once and has never so fully known it since,"[35] and not just a study but an actual entrance into it, only as a tourist, perhaps, but an entrance. The transition is a spatial one:

All this time we have been standing on the *parvis,* looking out over the sea and sands which are as good eleventh-century landscape as they ever were; or turning at times towards the church door which is the *pons seclorum,* the bridge of ages, between us and our ancestors. Now that we have made an attempt, such as it is, to get our minds into a condition to cross the bridge without breaking down in the effort, we enter the church. . . . Serious and simple to excess! is it not? Young people rarely enjoy it. They prefer the gothic, even as you see it here, looking at us from the choir, through the great Norman arch. No doubt, they are right, since they are young: but men and women who have lived long and are tired,—who want rest,—who have done with aspirations and ambition,—whose life has been a broken arch—feel this repose and self-restraint as they feel nothing else.[36]

What Adams seeks in this spatial world is a sense of unity and completeness, an ability to reconcile all the opposites of the world into a single, atemporal comprehension, which his own book mimes:

Barring her family quarrels, Europe was a unity then, in thought, will and object. Christianity was the unit. Mont Saint Michel and Byzantium were near each other. The Emperor Constantine and the Emperor Charlemagne were figured as allies and friends in the popular legend.... The whole Mount still kept the grand style; it expressed the unity of Church and State, God and Man, Peace and War, Life and Death, Good and Bad; it solved the whole problem of the universe.... The world is an evident, obvious, sacred harmony.... One looks back on it all as a picture; a symbol of unity; an assertion of God and Man in a bolder, stronger, closer union than ever was expressed by other art.[37]

This is Dante's unity, to depart from which is his definition of sin; it is also Nietzsche's life within horizons, his unconditional love, which he can only imagine as a self-willed illusion, a deliberate assertion contrary to truth. It is here conceived of, however, as the truth, eternally comprehensible, never to be superseded. Was it only an elaborate illusion, believed without the mental horror of Nietzsche's self-consciousness? The uncle is careful not to say, seemingly expresses seeming belief without commitment. What is certain is that he finds the image of unity entirely beautiful, understands it, wishes he could inhabit it, tourist as he is. At the center of the unity is its empowering force, the Virgin. Her power is love and generation; she is the God-bearer, the mother of the world, patroness of all conception and birth, of all acts of love and cooperation that allow that condition of love to exist which Augustine called a city. She is the meek submitter to God's necessity, who is raised to the greatest act of power. She reconciles the male with the female. The architectural figure of her worship is the arch, the spatial reconciliation of the horizontal with the vertical aspiration toward God, expressed in perfect symmetry. The arch finds its limit in the broken arch, "the finite idea of space":[38]

Art is a fairly large field where no one need jostle his neighbor, and no one need shut himself up in a corner; but, if one insists on taking a corner of preference, one might offer some excuse for choosing the Gothic Transition. The quiet, restrained strength of the romanesque married to the graceful curves and vaulting imagination of the gothic makes a union nearer the ideal than is often allowed in marriage. The French, in their best days, loved it with a constancy that has thrown a sort of aureole over their fickleness since. They never tired of its possibilities. Sometimes they put the pointed arch within the round, or above it; sometimes they put the round within the pointed. Sometimes a roman arch covered a cluster of pointed windows, as though protecting and caressing its children; sometimes a huge pointed arch covered a great rose-window spreading across the whole front of an enormous cathedral, with an arcade of romanesque windows beneath.

The French architects felt no discord, and there was none. Even the pure gothic was put side by side with the pure roman. You will see no later gothic than the choir of the Abbey Church above (1450–1521), unless it is the north *flèche* of Chartres cathedral (1507–1513); and if you will look down the nave, through the triumphal arches, into the pointed choir four hundred years more modern, you can judge whether there is any real discord. For those who feel the art, there is none; the strength and the grace join hands; the man and woman love each other still . . . no architecture that ever grew on earth except the gothic, gave this effect of flinging its passion against the sky.

When men no longer felt the passion, they fell back on themselves, or lower. The architects returned to the round arch, and even further to the flatness of the Greek colonnade; but this was not the fault of the twelfth or thirteenth centuries. What they had to say they said; what they felt they expressed; and if the seventeenth century forgot it, the twentieth in turn has forgotten the seventeenth. History is only a catalogue of the forgotten.[39]

The spatial world of art and architecture and intellection, in which all is harmonized, is contrasted to the temporal world of history, that "catalogue of the forgotten," in which nothing can be connected, which, in its supersessive forgettings of what each past generation thought to be true, finds its true subject to be illusion:

> . . . you can lay the history of the matter on the shelf for study at your leisure, if you ever care to study into the weary details of human illusions and disappointments, while here we pray to the Virgin, and absorb ourselves in the art, which is your pleasure and which shall not teach either a moral or a useful lesson.[40]

The sleight of logic that Adams performs is simple and elegant and devastating. Nietzsche saw scientific history as factual, solid, true, destroying the illusions of humankind by revealing them in their relativistic multiplicity, transforming all attempts at truth into a world of competing illusions, a world anticipated by Abelard's opposition of the authorities in *Sic et non*. Adams robs history of all its seeming solidity and authority by revealing its subject to be illusory. If supersessive history is true in its premises then all that it treats of in the past becomes insubstantial. History is a history of illusions superseding one another, a science of nothing, not even of anything as substantial as memory, but of forgetting. And in contrast to this aery science, the art and architecture, the products of illusion, take on immense, powerful, and simultaneous substance.

The structure of *Mont Saint Michel and Chartres* is fugal as well as spatial. Architecture provides the grounding theme, which then expands to encompass the poetry, the theology, the philosophy, the whole culture of the Middle Ages, all considered spatially, that is, simultaneously. As more is added, nothing is abandoned, all accumulates into greater and more complex symmetry, as the architecture of Aquinas's philosophy balances the architecture of the cathedral. Submerged beneath all is the hint of a countertheme of time, as the uncle's discourse is forced, reluctantly, from the exposition of simultaneous space and form, to the transition of the romanesque to the gothic, the latter being the architecture of supreme peril, aspiring most toward heaven in its pointed and broken arch, but also elaborating the limits of material and conception, the possibility of fall, both of the physical church and of Aquinas's Church. The book ends with that peril at the ultimate point of achievement and danger, in a fugue of rhetoric that redoubles and contains all the themes of the book, ending with a sudden and astonishingly final closure:

> The pathetic interest of the drama deepens with every new expression, but at least you can learn from it that your parents in the nineteenth century were not to blame for losing the sense of unity in art. As early as the fourteenth century, signs of unsteadiness appeared, and, before the eighteenth century, unity became only a reminiscence. The old habit of centralizing a strain at one point, and then dividing and subdividing it, and distributing it on visible lines of support to a visible foundation, disappeared in architecture soon after 1500, but lingered in theology two centuries longer, and even, in very old-fashioned communities, far down to our own time; but its values were forgotten, and it survived chiefly as a stock jest against the clergy. The passage between the two epochs is as beautiful as the Slave of Michael Angelo; but, to feel its beauty, you should see it from above, as it came from its radiant source. Truth, indeed, may not exist; science avers it to be only a relation; but what men took for truth stares one everywhere in the eye and begs for sympathy. The architects of the twelfth and thirteenth centuries took the Church and the Universe for truths, and tried to express them in a structure that should be final. Knowing by an enormous experience precisely where the strains were to come, they enlarged their scale to the utmost point of material endurance, lightening the load and distributing the burden until the gutters and gargoyles that seem mere ornament, and the grotesques that seem rude absurdities, all do work either for the arch or for the eye; and every inch of material, up and down, from crypt to vault, from Man to God, for the Universe to the Atom, had its task, giving support where support was needed, or weight where concentration

was felt, but always with the condition of showing the great lines which lead to unity and the curves which controlled divergence; so that, from the cross on the *flèche* and the key-stone of the vault, down through the ribbed *nervures*, the columns, the windows, to the foundation of the flying buttresses far beyond the walls, one idea controlled every line; and this is true of Saint Thomas's Church as it is of Amiens Cathedral. The method was the same for both, and the result was an art marked by singular unity, which endured and served its purpose until man changed his attitude towards the universe. The trouble was not in the art or the method or the structure, but in the universe itself which presented different aspects as man moved. Granted a Church, Saint Thomas's Church was the most expressive that man has made, and the great gothic Cathedrals were its most complete expression.

Perhaps the best proof of it is their apparent instability. Of all the elaborate symbolism which has been suggested for the gothic Cathedral, the most vital and most perfect may be that the slender *nervure*, the springing motion of the broken arch, the leap downwards of the flying buttress,—the visible effort to throw off a visible strain,—never let us forget that Faith alone supports it, and that, if Faith fails, Heaven is lost. The equilibrium is visibly delicate beyond the line of safety; danger lurks in every stone. The peril of the heavy tower, of the restless vault, of the vagrant buttress; the uncertainty of logic, the inequalities of the syllogism, the irregularities of the mental mirror,—all these haunting nightmares of the Church are expressed as strongly by the gothic Cathedral as though it had been the cry of human suffering, and as no emotion had ever been expressed before or is likely to find expression again. The delight of its aspirations is flung up to the sky. The pathos of its self-distrust and anguish of doubt, is buried in the earth as its last secret. You can read out of it whatever else pleases your youth and confidence; to me, this is all.[41]

These last pages contain the whole work.

The right panel of the diptych could not be more different from its counterpart. Where *Mont Saint Michel* reveals a growing and finally absolute identification with its subject, *The Education of Henry Adams* possesses no identification at all. The disembodied narrator— there is no equivalent to the relationship of uncle/author to niece/ reader—speaks of Adams always in the third person. Adams is merely a form, a tailor's manikin to be dressed and undressed in the supersessive fashions of history, philosophy and ideology (the image is of course borrowed from Carlyle's *Sartor Resartus*):

The manikin, therefore, has the same value as any other geometrical figure of three or more dimensions, which is used for the study of relation. For that purpose it cannot be spared; it is the only measure of motion, of proportion, of human condition; it must have the air of reality; must be taken for real; must be treated as though it had life;—Who knows? Possibly it had![42]

A study in medieval architecture involves the sympathetic identification of the writer; his own autobiography evokes an utter detachment.

Where *Mont Saint Michel* is spatial in its structure, exfoliating outward in the simultaneous complexity of architecture, adding always by accumulation while abandoning nothing, *The Education* has an entirely temporal and linear structure. Its progress is chronological; the clothes are assumed and discarded; it follows a single path of biography. The image of the road or path recurs throughout, often as a severed road: "but Mount Vernon always remained where it was, with no practical road to reach it," or "On that line, too, education could not go further. Tammany Hall stood at the end of the vista."[43] The linear road, cut in so many places, requiring so many changes of direction, so many fresh starts, becomes at last a maze, a catacomb of history:

> Yet the search for a unit of force led into catacombs of thought where hundreds of thousands of educations had found their end. Generation after generation of painful and honest-minded scholars had been content to stay in these labyrinths forever, pursuing ignorance in silence, in company with the most famous teachers of all time. Not one of them had ever found a logical highroad of escape.[44]

The labyrinth is still a linearity; it does not possess the simultaneous complexity of the cathedral. The characteristic of the labyrinth is that one can comprehend only a tiny segment of it at a time (this is what being lost means, whether temporally or spatially), divorced from all other segments by turns before and after. This is Adams's image of history; it is T. S. Eliot's history of "many cunning passages, contrived corridors,"[45] a complexity that can only be experienced sequentially, never in simultaneity.

Within the structure of road and labyrinth, the manikin Adams pursues his quest of an education until his biography becomes transmuted into a theory or antitheory of history, a vision of his age. Born the scion of two presidents, in Boston by the State House, into the Unitarian religion, into an intellectual world more of the eighteenth cen-

tury than of the nineteenth, he is propelled by the necessity of his time
toward the twentieth, in quest of an education. The essential quality
of an education is that it should possess completeness: "Any intelligent
education ought to end when it is complete."[46] He seeks "some great
generalization which would finish one's clamour to be educated."[47]
The path of education leads through Harvard College, and seems to
point toward Emerson, "the Concord faith,"[48] but this proves to be
another severed road:

> He never reached Concord, and to Concord Church, he, like the rest of
> mankind, who accepted a material universe, remained always an insect, or
> something much lower,—a man. It was surely no fault of his that the uni-
> verse seemed to him real; perhaps,—as Mr. Emerson justly said,—it was
> so; in spite of the long-continued effort of a life-time, he perpetually fell
> back into the heresy that if anything universal was unreal, it was himself
> and not the appearances; it was the poet and not the banker; it was his own
> thought, not the thing that moved it.[49]

Adams is not capable of the extreme subjectivist assertion of transcendentalism, and, in a mental world governed by a supersessive past,
such an assertion-against-truth is the only way to stop the process, to
circumscribe life, and education, within horizons. Adams is condemned, therefore, to process, to incompletion of knowledge in quest
of completion of knowledge. The quest leads him to politics, to the
academy, throughout Europe in a retracing of America's cultural and
temporal displacement, to Hegel and Marx and Darwin. Darwin
attracts him greatly: here seems to be a principle to explain change,
but all the instances given to him of uniformitarianism and of natural
selection in the course of many discussions with Sir Charles Lyell seem
to him dubious. The ice ages seem to him more catastrophic than uniform, and the case of *Terebratula* and other forms that survive
unchanged from the earliest fossils to the present strikes him as "altogether too much uniformity and much too little selection."[50] He writes
of the view from Wenlock Edge, with its thirteenth-century abbey and
fifteenth-century prior's house, haunts of his in the summer of the discussions with Lyell:

> The peculiar flavor of the scenery had something to do with absence of
> evolution; it was better marked in Egypt; it was felt wherever time-
> sequences became interchangeable. One's instinct abhors time. As one lay
> on the slope of the Edge, looking sleepily through the summer haze towards

Shrewsbury or Cader Idris or Caer Caradoc or Uriconium, nothing suggested sequence. . . . He could detect no more evolution in life since the *Pteraspis* than he could detect it in architecture since the Abbey. All he could prove was change . . . he could prove only Evolution that did not evolve; Uniformity that was not uniform; and Selection that did not select. To other Darwinians—except Darwin—Natural Selection seemed a dogma to be put in the place of the Athanasian creed; it was a form of religious hope; a promise of ultimate perfection. Adams wished no better; he warmly sympathised in the object; but when he came to ask himself what he truly thought, he felt that he had no Faith; that whenever the next new hobby should be brought out, he would surely drop off from Darwinism like a monkey from a perch; that the idea of one Form, Law, Order or Sequence, had no more value for him than the idea of none; that what he valued most was Motion, and that what attracted his mind was Change.[51]

From this point on the mind of the manikin Adams is seduced increasingly by the idea of motion and of change without form or purpose, an idea of supersession that makes any comprehension or completion impossible, in which no statement can be made to stay in place, in which, in Marx's phrase, "everything seems pregnant with its contrary."[52] The process continues through his years as professor of history at Harvard, searching for a valid and scientific methodology, and is hastened by his personal hauntings: the Civil War, which remains for him always an indicator of the impossibility of action because of the impossibility of completing knowledge, and action, however well motivated, stemming from incomplete knowledge has unforeseeable and catastrophic results; the death of his sister from tetanus, which convinces him of the hostility of the natural world;[53] the death by suicide of his wife, which is almost entirely repressed from the book, mentioned only by means of a description of her funerary monument.[54] The end of the process is a complete renunciation of the whole matter and purpose of history, including a renunciation of his own work:

Historians undertake to arrange sequences,—called stories, or histories,— assuming in silence a relation of cause and effect. These assumptions, hidden in the depths of dusty libraries, have been astounding, but commonly unconscious and childlike; so much so, that if any captious critic were to drag them to light, historians would probably reply, with one voice, that they had never supposed themselves required to know what they were talking about. Adams, for one, had toiled in vain to find out what he meant.

He had even published a dozen volumes of American history for no other purpose than to satisfy himself whether, by the severest process of stating, with the least possible comment, such facts as seemed sure, in such order as seemed rigorously consequent, he could fix for a familiar moment a necessary sequence of human movement. . . . Satisfied that the sequence of men led to nothing and that the sequence of their society could lead no further, while the mere sequence of time was artificial, and the sequence of thought was chaos, he turned at last to the sequence of force; and thus it happened that, after ten years' pursuit, he found himself lying in the Gallery of Machines at the Great Exposition of 1900, with his historical neck broken by the sudden irruption of force totally new.[55]

Adams, convinced now that the rigorous reductionism—the flight from narrative that seeks some statement of truth by reducing the definition of truth—of scientific history has led to no truth, but only to a reduction of ambition, finds new motive for history in increasing supersession, increasing the speed of history that makes all things unknowable, in propelling himself and history toward the apocalypse or the abyss:

Adams proclaimed that in the last synthesis, order and anarchy were one, but that the unity was chaos. As anarchist, conservative and christian, he had no motive or duty but to attain the end; and, to hasten it, he was bound to accelerate progress; to concentrate energy; to accumulate power; to multiply and intensify forces; to reduce friction, increase velocity and magnify momentum, partly because this was the mechanical law of the universe as science explained it; but partly also in order to get done with the present which artists and some others complained of; and finally,—and chiefly—because a rigorous philosophy required it, in order to penetrate the beyond, and satisfy man's destiny by reaching the largest synthesis in its ultimate contradiction.[56]

By christian, he means a rather etiolated cultural christian,[57] the product of an "eighteenth-century education when God was a father and nature a mother, and all was for the best in a scientific universe."[58] To the end this is his aesthetic preference; the dynamism and endless, uncontrollable multiplicity he proclaims in the new twentieth-century "multiverse"[59] he hates, but it is this that draws his mind. The attraction leads in two directions: backward to a Christian unity more ancient and powerful than his eighteenth-century Unitarianism, and forward into a history speeding out of control, "an acceleration . . . of vertiginous violence"[60] leading into the "supersensual chaos."[61] He conceives of the two, in reverse order of chronology, as "The Dynamo and the Virgin."[62] The Virgin is the image of force for the twelfth century. This is "Force as Love." Her force is one of union with the Son in love, a force of natural procreation and union:

> This is the Church of Christ! If you seek him through me, you are welcome, sinner or saint; but he and I are one. We are Love! We have little or nothing to do with God's other energies which are infinite.[63]

The course of history since is, for Adams, not so much a secularization, but an abandonment of the Virgin and the Son, the manifestations of love and Incarnation, the humility of God, those aspects of God that are the province of the Church, and a pursuit instead of the Father, of the power and infinity that were never the proper province of human knowledge or emulation. The quest is still for God, but for God as the embodiment of omnipotence and omniscience. The best exposition of this idea is Adams's poem "Prayer to the Virgin of Chartres," in which he specifically equates the rejection of God's humility, the Virgin and the Son, with the migration to America:

> Crossing the hostile sea, our greedy band
> Saw rising hills and forests in the blue;
> Our father's kingdom in the promised land!
> —We seized it, and dethroned the father too.
>
> And now we are the Father, with our brood,
> Ruling the Infinite, not Three but One;
> We made our world and saw that it was good;
> Ourselves we worship, and we have no Son.[64]

Adams's attraction to the Virgin, the expression of which became *Mont Saint Michel and Chartres*, becomes for him his love, a love of Love as the intellectual principle of the universe. The writing and contemplation of *Mont Saint Michel and Chartres* is a very real encounter and very real threat to Adams:

> This education startled even a man who had dabbled in fifty educations all over the world; for, if he were obliged to insist on a Universe, he seemed driven to the Church. Modern science guaranteed no unity. The student seemed to feel himself, like all his predecessors, caught, trapped, meshed in this eternal drag-net of religion.[65]

But the dragnet, both promise and threat, does not hold him; Adams shares with Nietzsche the peculiar modern sensibility of believing those things he loves to be illusion and those things he hates to be true. *Mont Saint Michel and Chartres* belongs to the past; it becomes, in the end, only a point from which to measure what has been lost, and Adams must return to twentieth-century multiplicity, to his own education, and the theory of history that concludes it.

This is the history of the dynamo, whose avatars Adams sees in the Gallery of Machines at the Paris Exposition of 1900. The dynamo is Adams's image of force for the twentieth century: impersonal, mechanistic, centrifugal, pragmatic. It is force measured by motion and speed. Adams proposes "A Dynamic Theory of History," and "A Law of Acceleration."[66] History is defined as the economy of forces, which economy manifests itself by change, motion, dynamism; and force is cumulative; the multiplication of force by force leads to an exponential acceleration of change throughout history, leading to supersensual chaos. Adams has chosen between the Universe and the Multiverse: "Chaos was the law of nature; Order was the dream of man."[67] With this choice and his formulation of the dynamic theory of history, "All the historian won was a vehement wish to escape. He saw his education complete, and was sorry he ever began it."[68] To the last, he is haunted by the vision of unity that provides the starting line, the vital measuring point, for his history:

> The stupendous failure of Christianity tortured history. The effort for Unity could not be a partial success; even alternating Unity resolved itself into meaningless motion at last.[69]

I have not tried to be sophisticated in my reading of Adams (he does not need it, for he is a writer of exposition and interpretation), but simply to present him, as much as possible in his own words. He exhibits, perhaps better than any other writer, and with an extraordinary poignancy, the sensation of living within supersessive history, at the threshold of the twentieth century. In a sense, history has not gone beyond Adams; his historical world is ours.[70]

Henry Adams sought to measure acceleration by fixing two historical points; I have sought to fill in the time between those two points, to provide a mechanism and a motive for the transition from a unifying, static, and traditionalist model of history, to a schismatic, dynamic, and revolutionary one—the historical paradigm that I have labeled supersessive—and to delineate, to some extent, the path of influence by textual demonstration.

In his preface to *Metahistory*, Hayden White writes that

> in any field of study not yet reduced (or elevated) to the status of a genuine science, thought remains the captive of the linguistic mode in which it seeks to grasp the outline of objects inhabiting its field of perception.[71]

One might question what a "genuine science" is, and whether it too is not the product of a linguistic mode. Apart from its excepting of science, the statement strikes me as eminently correct, as does the whole of White's book, except that it does not go far enough. The problem lies in the phrase "outline of objects inhabiting its field of perception." In the case of history, there are no objects inhabiting the field of perception. The past is, by its very pastness, removed from the instruments of empirical observation. When dealing with the past, we cannot escape the question posed by Augustine in the brilliant and unsurpassed philosophical meditation on the nature of memory and time that concludes the *Confessions*. I have described this in some detail in the first chapter; it can be paraphrased and concentrated in the single question: Where is this thing that we call past time? The answer cannot be given materialistically; archaeology provides only a few points from which to reconstruct the pointillist picture of the lost past; it will not provide us with narrative, and narrative is what we crave; philosophically it will not help at all: Constantine's dust is not Constantine. The past exists in memory and in communal memory, expressed in literary texts. It will not do to deconstruct these texts, as Bultmann did the Gospels, in the hope of finding something "real" or "actual" behind them, for this can only be a figment of imagination in spite of the evidence rather than in accord with it. The objects that occupy the historical field of vision are precisely texts, histories, literary accounts of the past imposed and superimposed upon one another like the writings of a gigantic palimpsest. As White has shown, history is a literary structure and the choices between various modes of historical imagination are aesthetic or moral choices, or even metaphysical choices.[72] This is not to say that all modes of history are equal discourses. They may be right or wrong in the same sense that aesthetic or moral statements may be right or wrong.

Once this essential literariness of history is accepted, then certain conclusions become inescapable from the material I have presented:

1. Both of the systems of history described in this study are responses to problems of earthly time and duration that are peculiar to Christian culture.

2. A change of historical system is preceded by a change in ideology. History is invented, although not consciously, to justify ideology. As an inverse correlative to this, a historical system will fail when it can no longer contain the phenomenal or conceptual world of its culture. Pagan history could not contain Christianity; the Eusebian/Augustinian/Orosian system of Christian history that superseded it could not contain Protestantism, nor a continual and indefinite expansion of time beyond the first millennium of the Church. It could probably, however, have contained Copernican cosmology and the New World.

3. The immediate causes of the two historical revolutions I have described were religious, not economic or scientific. In both cases the culture was possessed by religious ideas that formed the primary conception of reality or truth. The sixteenth and seventeenth centuries were racked by religious wars, not by cosmological wars of Copernicans against Ptolemaicists. The advent of modern scientific conceptions is contributory but not primary as a causative agent in the formation of the new history; it is more largely derivative of the new history; it could not have developed without the historical idea of the new and without an imperative to develop a conception of true knowledge, "science," apart from the traditional and religious bases of knowledge.

4. Although history is in some sense "invented" as an adjunct to ideology, it also exerts a tremendous coercive force on ideology and philosophy. Only an ideological upheaval is sufficiently powerful, usually in combination with many other factors, to break the image of the past. A history tends to force a culture that holds it to follow its logical consequences to their limits. Nietzsche and Adams both exhibit the difficulties of thinking outside the terms of one's inherited historical system, or even of describing it from within, even when such a desire exists.

5. The Reformation and Renaissance historical revolution was in effect a complete and relatively sudden reversal of the Eusebian/ Augustinian/Orosian system. It exploited and made manifest the suppressed material of that system: the possibility of dynamic mutation that would make adherence to an original truth impossible.

6. The Eusebian/Augustinian/Orosian history tended to the erection of a unified and complete conception of the world, to faith. It is constructive and, therefore, as Adams understood in his image of the cathedral, fragile. It has a very limited tolerance for the new or for dissent and, among its many very real virtues, it tends to the vice of spiritual totalitarianism.

7. Supersessive history is, by contrast, inherently deconstructive in nature. It tends to foster skepticism and multiplicity, as it historicizes (that is, temporalizes and relativizes) all ideologies. Indeed, the term ideology in its modern usage—a systematic statement of truth conceived of apart from truth itself, that is, as a discourse—is the product of and is inconceivable apart from this historical mode. Because of its supersessive form, it has an infinite tolerance for the new, for successive waves of revolution. Also because it builds no system other than the supersession of systems, it is almost impervious to destruction. For this reason, the recent rebellion against historical consciousness in Sartre, Lévi-Strauss, Foucault, and others[73] may not be a counter to the supersessive system of history, but merely a new revolution within it, and controlled by its structure. Certainly this movement uses deconstructive and skeptical methodology which seems to be inherent to supersessive history from its inception (one thinks of Luther's imitation of Valla).

8. The most distinctive cultural characteristic of supersessive history is its contempt for the past. Wittgenstein noted in his journals that

> um einen Dichter zu geniesen, dazu mus man auch die Kultur, zu der er gehört, *gern haben*. Ist die einem gleichgültig oder zuwider, so erkaltet die Bewunderung.

> Eine Zeit misversteht die andere; und eine *kleine* Zeit misversteht alle andern in ihrer eigenen häslichen Weise.

> (if one is to enjoy a writer one has to *like* the culture he belongs to as well. If one finds it indifferent or distasteful, one's admiration cools off.

One age misunderstands another; and a *petty* age misunderstands all the others in its own nasty way.)[74]

Supersessive history gives rise to a succession of such petty ages, each in ideological conflict with all of the others. Although there are many individual and communal exceptions involving idealizations of past culture—but never held in synthesis, always in opposition to the present or to the immediate past—the general tone of historical scholarship from Valla to the present has been one of dismissive contempt for, because of ideological opposition to, its materials.

9. Because of the necessary relativism of supersessive culture, it tends to foster either extreme subjective assertion or ever-increasing skepticism.[75] The first of these is a periodic phenomenon; the latter tends to increase as the past history and present society multiply competing ideologies until the faintest remnants of the ideal of unity in objective truth are lost. The two are aspects of each other. They are both results of ideology conceived in conflict rather than in continuum, in schism rather than in syncretism, with the past and with other contemporaneous ideologies. Of colonialism and of Western missionary enterprise, conceived of as ideological colonization, Simone Weil wrote that they had spread Western unbelief like a contagion, that we of the West, who believe almost nothing, have created everywhere those who believe nothing at all.[76]

Notes

Preface

1. Perry Miller and Thomas H. Johnson, *The Puritans: A Sourcebook of Their Writings*, 2 vols. (New York: Harper, 1963) 1:291.

2. Thomas Hooker, *A Survey of the Summe of Church-Discipline* . . . (London, 1648) A2 recto; also cited in Miller and Johnson 1:85.

3. Francis Bacon uses the phrase in its original sense (*Novum Organum* bk. 1, aphorism 84), with complex reinterpretations of his own. According to R. G. Collingwood (*The Idea of History* [1946; London: Oxford University Press, 1956] 72), Bacon derives the phrase from *Noctes Atticae* of Aulus Gellius, although the attribution is not in Bacon.

4. Michel de Certeau, *L'Écriture de l'histoire* (Paris: Gallimard, 1975); *The Writing of History*, trans. Tom Conley (New York: Columbia University Press, 1988); Hayden White, *Metahistory: The Historical Imagination in Nineteenth-Century Europe* (Baltimore: Johns Hopkins University Press, 1973); *The Content of the Form: Narrative Discourse and Historical Representation* (Baltimore: Johns Hopkins University Press, 1987); Paul Ricoeur, *Histoire et vérité*, 2nd ed. (Paris: Seuil, 1964); *History and Truth: Essays*, trans. Charles A. Kelbley, Northwestern University Studies in Phenomenology and Existential Philosophy (Evanston, Ill.: Northwestern University Press, 1966); *Temps et récit*, 3 vols. (Paris: Seuil, 1983–1985); *Time and Narrative*, trans. Kathleen McLaughlin and David Pellaur, 3 vols. (Chicago: University of Chicago Press, 1984–1988); Heinrich Rickert, *Die Grenzen der naturwissenschaftlichen Begriffsbildung: Eine logische Einleitung in die historische Wissenschaften*, 5th ed. (Tübingen: Mohr, 1929); *The Limits of Concept Formation in Natural Science: A Logical Introduction to the Historical Sciences*, ed. and trans. Guy Oakes, Texts in German Philosophy (Cambridge: Cambridge University Press, 1986); R. G. Collingwood, *The Idea of History*, (1946; London: Oxford University Press, 1956). See also Albert Cook, *History/Writing* (Cambridge: Cambridge University Press, 1988); Arthur C. Danto, *Narration and Knowledge* (New York: Columbia University Press, 1985); W. B. Gallie, *Philosophy and*

the Historical Understanding, 2nd ed. (New York: Schocken, 1968); J. L. Gorman, *The Expression of Historical Knowledge* (Edinburgh: Edinburgh University Press, 1982); David Lowenthal, *The Past Is a Foreign Country* (Cambridge: Cambridge University Press, 1985).

5. Maurice Mandelbaum, *The Problem of Historical Knowledge: An Answer to Relativism* (New York: Liveright, 1938); *The Anatomy of Historical Knowledge* (Baltimore: Johns Hopkins University Press, 1977); Leon J. Goldstein, *Historical Knowing* (Austin: University of Texas Press, 1976).

6. J. B. Bury, *The Idea of Progress: An Inquiry into Its Origin and Growth* (New York: Macmillan, 1932); Reinhart Koselleck, *Kritik und Krise: Eine Studie zur Pathogenese der bürgerlichen Welt* (Freiburg: Karl Alber, 1959); *Critique and Crisis: Enlightenment and the Pathogenesis of Modern Society* (Cambridge, Mass.: MIT Press, 1988).

Chapter 1

1. For an excellent overview of the early texts of Christian history, their importance and influence, see Indrikis Sterns, *The Greater Medieval Historians: An Interpretation and Bibliography* (Lanham, Md.: University Press of America, 1980) 1–40.

2. Sterns 6–8, 24, 87.

3. John 14:15–17, 25–26; 15:26–27; 16:7–11, 13–14.

4. Matt. 24:3–46; Mark 13:3–37; Luke 12:35–40; 21:10–36.

5. I use the names of authors to denote the texts attached to them (the idea of authorship) rather than any historical persons.

6. 1 John 4:1. In all New Testament citations, Greek text is from Kurt Aland, et al., eds., *The Greek New Testament*, 3rd ed. (New York: American Bible Society, 1975); translation is from the Revised Standard Version (RSV).

7. Eusebius of Caesarea, *Eusebius: The Ecclesiastical History*, trans. Kirsop Lake, 2 vols., Loeb Classical Library (London: Heinemann; New York: Putnam's, 1926), 1:256–59; bk. 3, chap. 25.

8. 1 John 2:18.

9. Eusebius 1:6–7; bk. 1, chap. 1.

10. Kirsop Lake, Introduction, *Eusebius: The Ecclesiastical History* 1:xxxiv–xxxv.

11. Eusebius 1:6–7; bk. 1, chap. 1.

12. Eusebius 1:4–5; bk. 1, Contents.

13. Eusebius 1:8–10; bk. 1, chap. 1.

14. Eusebius 1:10–11; bk. l, chap. 1.

15. Eusebius 1:16–17; bk. 1, chap. 1.

16. Eusebius 1:276–77; bk. 3, chap. 32.

17. Eusebius 1:280–81; bk. 3, chap. 36.

18. See, for instance, Eusebius 1:293–97; bk. 3, chap. 39.

19. Michel Foucault, *L'Archéologie du savoir* (Paris: Gallimard, 1969) 22–23; *The Archaeology of Knowledge and the Discourse on Language*, trans. A. M. Sheridan Smith (New York: Pantheon, 1982) 12.

20. Eusebius 1:110–13 and notes; bk. 2, chap. 2.

21. See, for instance, Eusebius 1:202–3; bk. 3, chaps. 5–6 and 1:214–15; bk. 3, chap. 7.

22. See, for instance, J. Gordon Melton, "Spiritualization and Reaffirmation: What Really Happens When Prophecy Fails," *American Studies* 26 (1985): 12–29.

23. 2 Thess. 2:1–12.

24. Matt. 24:27.

25. 1 John 3:18–19.

26. William Green, Introduction, *Saint Augustine: The City of God Against the Pagans*, trans. Eva Matthews Sanford and William Green, 7 vols., Loeb Classical Library (Cambridge, Mass.: Harvard University Press; London: Heinemann, 1957–72) 5:xiv–xv. Subsequent references are to this edition; translator varies by volume and is given at first citation of each.

27. Augustine, *The City of God*, trans. Philip Levene, 4:62–64; bk. 12, chap. 14.

28. Augustine, *The City of God* 4:44–109; bk. 12, chaps. 10–21.

29. Augustine, *The City of God* 4:460–69; bk. 15, chaps. 10–11.

30. In *A Select Library of Nicene and Post-Nicene Fathers of the Christian Church*, ed. Philip Schaff and Henry Wace, 1st ser., 14 vols. (New York, 1886–1900) 8: Psalm 93.

31. Augustine, *The City of God*, trans. Philip Levene (4), E. M. Sanford and W. Green (5), W. Green (6), 4:2 to 6:453; bk. 12, chap. 1 to bk. 20, chap. 30.

32. Augustine, *The City of God* 5:202–5; bk. 16, chap. 43.

33. Augustine, *The City of God* 6:286–87; bk. 20, chap. 7.

34. Bk. 11, chap. 10–29.

35. Augustine, *Saint Augustine's Confessions*, trans. William Watts (1631), 2 vols., Loeb Classical Library (Cambridge, Mass.: Harvard University Press; London: Heinemann, 1950–1951) 2:240–41: bk. 11, chap. 15.

36. Augustine, *Confessions* 2:240–43; bk. 11, chap. 15.

37. Augustine, *Confessions* 2:246–49, 274–75, 276–77; bk. 11, chap. 17, 18, 27, 28.

38. Augustine, *The City of God*, trans. David Wiesen, 3:140–41; bk. 8, chap. 27.

39. Augustine, *The City of God*, trans. William Greene, 7:208–45; bk. 22, chap. 8. Cf. Peter Brown, *The World of Late Antiquity, AD 150–750*, History of European Civilization Library (New York: Harcourt, 1971) 181–83.

40. Tertullian, *Apologeticus*, in *Tertullian: Apology, De spectaculis; Minucius Felix*, trans. T. R. Glover and Gerald H. Rendall, Loeb Classical Library (London: Heinemann; New York: Putnam's, 1931) 204–05; chap. 46.

41. Augustine, *The City of God* 6:236–37; bk. 19, chap. 26.

42. Augustine, *The City of God* 4:258–61; bk. 14, chap. 1.

43. Augustine, *The City of God* 6:232–33, bk. 19, chap. 24.

44. Augustine, *The City of God* 4:404–7; bk. 14, chap. 28; 4:422–25; bk. 15, chap. 4; 6:236–37; bk. 19, chap. 26.

45. Augustine, *The City of God* 4:418–19; bk. 15, chap. 2.

46. Augustine, *The City of God* 6:282–85; bk. 20, chap. 7.

47. Augustine, *The City of God* 6:286–313; bk. 20, chaps. 7–9.

48. Rev. 13–18.

49. Augustine, *The City of God* 6:326–27; bk. 20, chap. 23.

50. Augustine, *The City of God* 6:312–13; bk. 20, chap. 19.

51. Rev. 13:11–18; 20:4–6.

52. Augustine, *The City of God* 6:312–13; bk. 20, chap. 19.

53. Brown 108.

54. Paulus Orosius, *Pauli Orosii historiarum adversum paganos libri VII*, ed. Carl Zangemeister (Leipzig, 1889); *Seven Books of History against the Pagans: The Apology of Paulus Orosius*, trans. Irving Woodworth Raymond, Records of Civilization: Sources and Studies (New York: Columbia University Press, 1936).

55. See C. A. Patrides, *The Grand Design of God: The Literary Form of the Christian View of History*, Ideas and Forms in English Literature (London: Routledge; Toronto: University of Toronto Press, 1972) 19–20; Ernst Breisach, *Historiography: Ancient, Medieval, and Modern* (Chicago: University of Chicago Press, 1983) 86; Irving Woodworth Raymond, Introduction, *Seven Books of History against the Pagans* 22.

56. See Raymond 20.

57. Bk. 7, chap. 28.

58. See Orosius's dedication.

59. Daniel 2:31–45.

Chapter 2

1. Bernard of Clairvaux, *Opera omnia*, ed. D. Joannis Mabillon, 4th ed., 2 vols. (Paris, 1839) 1:389–90 (letter 174); *The Letters of Saint Bernard of Clairvaux*, trans. Bruno Scott James (Chicago: Henry Regnery, 1953) 289–90 (letter 215).

2. Bernard of Clairvaux, *Opera omnia* 1:393; *Letters* 293.

3. Bernard of Clairvaux, *Opera omnia* 1:412 (letter 189); *Letters* 318 (letter 239).

4. Henry VIII, *Assertio septem sacramentorum adversus Martin Lutherus* ([London and Rome], 1521) n. pag.; *Assertio septem sacramentorum: or, An Assertion of the Seven Sacraments against Martin Luther*, trans. T[homas]. W[ebster]. (London, 1687) n. pag. (Preface), 41, 47.

5. Henry VIII, *Assertio* n. pag.; *Assertio* (Webster's trans.) 101.

6. Gregory of Tours, *Gregoire de Tours: Histoire des Francs*, ed. René Poupardin, Collection de textes pour servir à l'étude et à l'enseignement de l'histoire (Paris: Alphonse Picard, 1913) 1; *The History of the Franks*, trans. Lewis Thorpe, Penguin Classics (Harmondsworth: Penguin, 1974) 63; Preface.

7. R. G. Collingwood, *The Idea of History* (1946; London: Oxford University Press, 1956) 55:

> Thus, in medieval thought, the complete opposition between the objective purpose of God and the subjective purpose of man, so conceived that God's purpose appears as the imposition of a certain objective plan upon history quite irrespective of man's subjective purposes, leads inevitably to the idea that man's purpose makes no difference to the course of history and that the only force that determines is the divine nature.... God in medieval theology is not substance but pure act; and transcendence means that the divine activity is conceived not as working in and through human activity but as working outside it and overruling it, not immanent in the world of human action but transcending that world.

8. Fulcher of Chartres, *Fulcheri Carnotensis historia Hierosolymitana*, ed. Heinrich Hagenmeyer (Heidelberg: Winters, 1913); *Fulcher of Chartres: Chronicle of the First Crusade*, trans. Martha Evelyn McGinty, Translations and Reprints from the Original Sources of History, 3rd ser. 1 (Philadelphia: University of Pennsylvania Press; London: Oxford University Press, 1941). Fulcher represents the height of optimism in any Crusader history; I offer him as the medieval writer who militates most against the thesis I am presenting here. For a darker, more typical history, one in which chaos and incoherence become the conscious subject, see William of Tyre, *Historia de rebus gestis in partibus transmarinis*, in *Guillaume de Tyr et ses continuateurs*, ed. Paulin Paris, 2 vols. (Paris, 1879); *A History of Deeds Done beyond the Sea*, trans. Emily Atwater Babcock and A. C. Krey, 2 vols, Records of Civilization: Sources and Studies (New York: Columbia University Press, 1943).

9. Matthew Paris, *Matthaei Parisiensis, monachi Sancti Albani, chronica majora*, ed. Henry Richards Luard, 7 vols., Chronicles and Memorials of Great Britain and Ireland during the Middle Ages (Rolls Series) 54 (London, 1877) 4:100; *Matthew Paris's English History*, trans. J. A. Giles, 3 vols. (London, 1852–1854) 1:332.

10. Dante Alighieri, *De monarchia*, ed. Karl Witte (Vienna, 1874) 93–95; *The De monarchia of Dante Alighieri*, trans. Aurelia Henry (Boston: Riverside-Houghton, 1904) 142–45; bk. 3, chap. 3.

11. The Monk of Saint Gall, *Monachus Sangallensis*, in *Monumenta Carolini*, ed. Philippe Jaffe, Bibliotheca rerum Germanicarum (Berlin, 1867) 639; *Early Lives of Charlemagne, by Eginhard and the Monk of St. Gall*, trans. and ed. A. J. Grant (London: Chatto, 1922) 72; bk. 1, chap. 10.

12. Paris, *English History* 1:24.

13. See, for instance, Paris, *English History* 1:97.

14. Bede, *Historia ecclesiastica gentis Anglorum*, in *Baedae opera historica*, trans. J. E. King, 2 vols., Loeb Classical Library (London: Heinemann; New York: Putnam's, 1930) 116–53; bk. 1, chap. 27.

15. Dante, *De monarchia* 30–31; *The* De monarchia *of Dante Alighieri* 54–56; bk. 1, chap. 15.

16. For a typical example, see C. D. Yonge, Translator's Preface, Matthew of Westminster, *The Flowers of History*, 2 vols. (London, 1853) 1:2–3.

17. See Matthew of Westminster, *Flores historiarum*, ed. Henry Richards Luard, 3 vols., Chronicles and Memorials of Great Britain and Ireland during the Middle Ages (Rolls Series) 95 (London, 1890) 1:108; *The Flowers of History*, trans. C. D. Yonge, 2 vols. (London, 1853) 1:137. Cf. Otto of Freising, *Ottonis episcopi Frisingensis chronica sive historia de duabus civitatibus*, ed. Adolf Hofmeister, Scriptores rerum Germanicarum in usum scholarum ex monumentis Germaniae historicis 45 (Hannover: Bibliopoli Hahniani, 1912) 146–47; *The Two Cities: A Chronicle of Universal History to the Year 1146* A.D., trans. Charles Christopher Mierow, Records of Civilization: Sources and Studies (New York: Columbia University Press, 1928) 235–36; bk. 3, chap. 11.

18. Otto of Freising, *Historia de duabus civitatibus* 75–78; *The Two Cities* 162–63; bk. 2, chap. 8. Cf. Augustine, *De civitate Dei* bk. 8, chap. 11, and his *De doctrina Christiana* bk. 2, chaps. 28 and 43.

19. Matthew of Westminster, *Flores Historiarum* 1:87–88; *Flowers of History* 1:117–18. It is interesting to compare this passage to the context given by Dante to the birth and death of Beatrice in *La Vita Nuova* chap. 29.

20. John Donne, "Hymn to God My God, in My Sickness."

21. George England and Alfred W. Pollard, eds., *The Townely Plays*, Early English Text Society, Extra Series 71 (1897; London: Oxford University Press, 1966) 261.

22. Robert de Boron, *Le Roman de l'estoire dou Graal*, ed. William A. Nitze (Paris: Champion, 1927).

23. See Augustine, *De civitate Dei* bk. 18, chap. 23; Otto of Freising, *Historia de duabus civitatibus* 72–73; *The Two Cities* 158; bk. 2, chap. 4; Matthew of Westminster, *The Flowers of History* 1:77 (omitted from *Flores historiarum* for editorial reasons, also in Matthew Paris).

24. See C. S. Lewis, *The Discarded Image: An Introduction to Medieval and Renaissance Literature* (Cambridge: Cambridge University Press, 1964) 1–12.

25. Geoffrey of Monmouth, *The* Historia Regum Brittanie *of Geoffrey of Monmouth*, ed. Acton Griscom (London: Longmans, 1929) 219; *History of the Kings of Britain*, trans. Sebastian Evans, rev. Charles W. Dunn (New York: Dutton, 1958) 3; bk. 1, chap. 1.

26. See Lucy Allen Paton, Introduction, Geoffrey of Monmouth, *History of the Kings of Britain* xvii–xx.

27. Chretien de Troyes, *Cligés*, ed. Wendelin Foerster, Christian von Troyes sämtliche Werke (Halle, 1884) 1–2; *Chretien de Troyes: Arthurian Romances*, trans. W. W. Comfort, ed. D. D. R. Owen, Everyman's Library (London: Dent; New York: Dutton, 1914) 91.

28. Bede 1–5; Preface.

29. Matthew of Westminster, *Flores historiarum* 1:1–2; *Flowers of History* 1:1–2; Preface.

30. George Santayana, *The Life of Reason: Or the Phases of Human Progress*, 5 vols. (New York: Scribner's, 1905–1906) 1 (*Introduction and Reason in Common Sense*): 284; chap. 12.

31. James Joyce, *Ulysses* (New York: Vintage-Random, 1961) 34: "—History, Stephen said, is a nightmare from which I am trying to awake."

32. Bede 1:84–87; bk. 1, chap. 18.

33. The following is an almost random selection of passages dealing with relics, chosen, if for any reason, for their typicality (because of their number, references are given in the most abbreviated form possible): *Gesta Francorum* last page (height and breadth of Christ from sepulcher); Matthew of Westminster (English ed.) 1:324 (Bones of Saint Benedict resurrect two corpses); 1:328 (Relics of Saint Sebastian deliver Rome from the plague); 1:364 (head of John the Baptist); Fulcher of Chartres chap. 18 (the lance); Matthew Paris (English ed.) 1:312 (crown of Christ); 1: 323 (the cross); William of Tyre bk. 1, chap. 1 (the cross); bk. 1, chap. 15 (Jerusalem a reliquary); Bede, *Historia Ecclesiastica* bk. 3, chap. 2 (King Oswald's cross).

34. For an example of the respect in which the Church was held, and the contempt in which the pope and prelates were held, and of the crimes of which they were accused, see the second tale of the first day in Boccaccio's *Decameron*.

35. Dante Alighieri, *De monarchia* 32–33, 37–38; *The* De monarchia *of Dante Alighieri* 59–61, 67–69; bk. 1, chap. 16; bk. 2, chap. 1.

36. Otto of Freising, *Historia de duabus civitatibus* 256; *The Two Cities* 353; bk. 5, chap. 31.

37. The years of "the City" provided the standard chronology of all Roman histories. Orosius also begins each of his chapters with the year of the city, and was followed in this by most medieval historians, resulting in an identification, or confusion, of Rome with Augustine's city of men. Interestingly, Otto has previously stated (bk. 3, chap. 6) that, after the birth of Christ he will use the Christian dating only, and no longer give the year of the city.

38. Otto of Freising, *Historia de duabus civitatibus* 142; *The Two Cities* 229–30; bk. 3, chap. 6.

39. See Helmold, *The Chronicle of the Slavs*, ed. Francis Joseph Tschan, Records of Civilization: Sources and Studies (New York: Columbia University Press, 1935) 116–20; chap. 33.

40. Otto of Freising, *Historia de duabus civitatibus* 1; *The Two Cities* 87; Dedication.

41. Charles Christopher Mierow, Introduction, Otto of Freising, *The Two Cities* 16.

42. Otto of Freising, *Historia de duabus civitatibus* 5; *The Two Cities* 91.

43. Otto of Freising, *Historia de duabus civitatibus* 6; *The Two Cities* 91.

44. Herman Melville, *Pierre: Or, the Ambiguities* bk. 14.

45. Otto of Freising, *Historia de duabus civitatibus* 6; *The Two Cities* 93.

46. See Mierow 28.

47. Otto of Freising, *Historia de duabus civitatibus* 8; *The Two Cities* 95.

48. Otto of Freising, *Historia de duabus civitatibus* 132; *The Two Cities* 220; bk. 3, Prologue.

49. See *Historia de duabus civitatibus* 141–43; *The Two Cities* 229–30; bk. 3, chap. 6. Augustus is identified as a type of Christ, and Christ was a Roman because of his enrollment in the census: Christ wished by this to show that he would make out of the city of the world, in some strange and inexpressible manner, his own city.

50. Otto of Freising, *Historia de duabus civitatibus* 192; *The Two Cities* 284; bk. 4, chap. 5.

51. Otto of Freising, *Historia de duabus civitatibus* 309; *The Two Cities* 404; bk. 7, Prologue.

52. Otto of Freising, *Historia de duabus civitatibus* 180–83; *The Two Cities* 271–74; bk. 4, Prologue.

53. Otto of Freising, *Historia de duabus civitatibus* 228; *The Two Cities* 323–24; bk. 5, Prologue.

54. Otto of Freising, *Historia de duabus civitatibus* 309; *The Two Cities* 404; bk. 7, Prologue.

Chapter 3

1. I have not quoted the passage because the whole canto has to be read in order to provide the necessary context.

2. For the most prominent examples of Wyclif's conflation of papacy and antichrist, see *Iohannis Wyclif: Tractatus de ecclesia*, ed. Johann Loserth (London, 1886) 356, 377, 451; *Iohannis Wyclif: Tractatus de apostasia*, ed. Michael Henry Dziewicki (London, 1888) 12, 46–47, 55, 149; *Iohannis Wyc-*

lif: Tractatus de blasphemia, ed. Michael Henry Dziewicki (London, 1893) 79, 106–8; *Johannis Wyclif: Tractatus de potestate pape,* ed. Johann Loserth (London: Wyclif Society-Trübner, 1907) 321–28.

3. See Dziewicki, Introduction, *Iohannis Wyclif: Tractatus de apostasia* xxv–xxvi.

4. Wyclif, *Tractatus de ecclesia* 8; chap. 1.

5. Wyclif, *Tractatus de ecclesia* 94; chap. 5.

6. Wyclif, *Tractatus de ecclesia* 366; chap. 16; my translation.

7. For questions of dating see Dziewicki, Introduction, *Iohannis Wyclif: Tractatus de blasphemia* vii–ix.

8. Wyclif, *Tractatus de potestate pape* 118; chap. 6, and elsewhere.

9. Wyclif, *Tractatus de potestate pape* 118–25; chap. 6.

10. Wyclif, *Tractatus de potestate pape* 322; chap. 12; my translation.

11. Daniel 11:31.

12. The whole argument is found in *Tractatus de potestate pape* 321–28; chap. 12.

13. Wyclif, *Tractatus de apostasia* 46; chap. 3; my translation.

14. Wyclif, *Tractatus de apostasia* 66; chap. 5; my translation.

15. Wyclif, *Tractatus de apostasia* 76–78; chap. 6.

16. See the textual apparatus of the several volumes.

17. See Loserth, Introduction, *Tractatus de potestate pape* xliv–xlvii for comparison of passages in *De potestate pape* and in Hus's *De ecclesia.* A collection of Hus's fragmentary writings on the antichrist formed the last volume of a collected edition of 1524: *Ioannes Huss de anatomia antichristi* (Strasbourg).

18. Martin Luther, *Supputatio annorum mundi,* 1541, Weimar 53:1–172. Note on Luther's texts: Latin or German text is from *D. Martin Luthers Werke,* 61 vols. (Weimar: Hermann Böhlau, 1883–1983), hereafter cited as Weimar, and English translation from *Luther's Works,* general eds. Jaroslav Pelikan and Helmut T. Lehman, 55 vols. (Saint Louis: Concordia, 1958–1968), hereafter cited as Saint Louis. Saint Louis by no means translates all of Weimar, and the greater part of Luther's work remains unavailable in English. Textual citations will comprise the title of the individual work in Weimar, the title in Saint Louis (if included) in parentheses, the date of the work (first citation only), volume and pages in Weimar, and in Saint Louis (if included). There are in most cases several editors/translators per work, and these will not be given.

19. Luther, *Supputatio annorum mundi,* Weimar 53:22.

20. Luther, *Supputatio annorum mundi,* Weimar 53:152–53.

21. Luther, *Supputatio annorum mundi,* Weimar 53:167.

22. Luther, *Von der Winkelmesse und Pfaffenweihe,* 1533, Weimar 38:251; trans. John M. Headley, *Luther's View of Church History,* Yale Publications

in Religion 6 (New Haven: Yale University Press, 1963) 222. Cf. Luther, *Operationes in Psalmos*, 1519–1521, Weimar 5:336–37, 339–40, 442, 479, 644, 649–50, 653, and *Wider das Papsttum zu Rom, vom Teufel gestiftet* (*Against the Roman Papacy, an Institution of the Devil*), 1545, Weimar 54:195–299; Saint Louis 41:263–376.

23. Luther, *Wider das Papsttum zu Rom, vom Teufel gestiftet* (*Against the Roman Papacy, an Institution of the Devil*), Weimar 54:232–35; Saint Louis 41:294–99.

24. Luther, *Vorlesung über I. Mose* (*Lectures on Genesis*), 1535–1545, Weimar 44:676; Saint Louis 8:132.

25. R. G. Collingwood, *The Idea of History* (1946; London: Oxford University Press, 1956) 42–45; Headley 56–94.

26. Luther, *Deuteronomion Mosi cum annotationibus* (*Lectures on Deuteronomy*), 1525, Weimar 14:690; Saint Louis 9:198–99. Cf. *Responsio ad condemnationem doctrinalem per Lovan. et Colon. factam*, 1520, Weimar 6:183.

27. Headley 25.

28. Luther, *Vorlesung über I. Mose* (*Lectures on Genesis*), Weimar 42:187; Saint Louis 1:252; *Predigten über das 2 Buch Mose*, 1524–1527, Weimar 16:17; *Operationes in Psalmos*, 1519–1521, Weimar 5:308–9; *Assertio omnium articulorum M. Lutheri per bullam Leonis X.*, 1520, Weimar 7:148.

29. Luther, *Wider das Papsttum zu Rom, vom Teufel gestiftet* (*Against the Roman Papacy, an Institution of the Devil*), Weimar 54:235; Saint Louis 41:298.

30. Luther, *Vorlesung über I. Mose* (*Lectures on Genesis*), Weimar 42:67; Saint Louis 1:88.

31. Luther, *Supputatio annorum mundi*, Weimar 53:152–53.

32. Dante Alighieri, *Purgatorio* 32: "... di quella Roma onde Cristo e romano."

33. See Headley 6.

34. Luther, *Epistola ad Romanos* (*Lectures on Romans*), 1515–1516, Weimar 56:446; Saint Louis 25:438–39.

35. Luther, *Predigten über das 2 Buch Mose*, Weimar 16:589; *Deuteronomion Mosi cum annotationibus* (*Lectures on Deuteronomy*), Weimar 14:602; Saint Louis 9:63.

36. Luther, *Responsio ad condemnationem doctrinalem per Lovan. et Colon. factam*, Weimar 6:183.

37. Luther, "Luther an Matthäus Rakeberger in Torgau," 25 March 1545, letter 4086 of *D. Martin Luthers Werke: Briefwechsel*, 18 vols. (Weimar: Hermann Böhlau, 1930–1985) 11:59; trans. Headley 265.

38. For general context and background of English Reformation history see F. J. Levy, *Tudor Historical Thought* (San Marino, Calif.: Huntingdon Library, 1967); F. Smith Fussner, *The Historical Revolution: English Histori-*

cal Writing and Thought, 1580–1640 (Westport, Conn.: Greenwood, 1962); Richard Bauckham, *Tudor Apocalypse: Sixteenth Century Apocalypticism, Millenarianism and the English Reformation: From John Bale to John Foxe and Thomas Brightman,* Courtney Library of Reformation Classics 8 (Oxford: Sutton Courtney, [1978?]).

39. John Foxe, *The Acts and Monuments of John Foxe,* 8 vols. (1870; New York: AMS, 1965). This is a reprint of the edition published in London by Seeley, ed. Josiah Pratt, which reproduces the text of the 1583 edition, Foxe's last revision.

40. Foxe 1: xxvii–xxxiv.

41. Foxe 1: xxxiv–xxxvi.

42. Foxe 1:2.

43. See George Townsend, "Life and Defence of John Foxe," 5–17. This biography is included in the first volume of the *Acts and Monuments,* with separate pagination. It is a quaint and biased work, which attributes all kinds of unknowable sentiments to its subject, but it is as good a compilation of the few available sources for and external facts of Foxe's life as any other.

44. John Foxe, *Commentarii rerum in Ecclesia gestarum . . .* (Strasbourg, 1554).

45. Foxe, *Acts and Monuments* 1:4.

46. Foxe 1:4–5.

47. Foxe 1:7, 9.

48. Foxe 1:89; bk. 1.

49. Foxe 1:292–304; bk. 1.

50. Foxe 1:300–3; bk. 1.

51. Foxe 1:310; bk. 2.

52. Foxe 1:112–13; bk. 1.

53. Foxe 1:258–59; bk. 1.

54. See Foxe's repetition of the argument at 1:300–3; bk. 1.

55. Luther, *Resolutio Lutheriana super propositione XIII. de potestate papae,* 1519, Weimar 2:209.

56. Luther, *De captivitate Babylonica ecclesiae praeludium* (*The Babylonian Captivity of the Church*), 1520, Weimar 6:561–62; Saint Louis 11:109.

57. Foxe 1:99; bk. 1.

58. Foxe 1:288–92; bk. 1.

59. Foxe 1:340–41; bk. 2.

60. Foxe 2:43, 64; bk. 3.

61. Foxe 2:47, 49; bk. 3.

62. Foxe 2:249; bk. 4.

63. Foxe 1:310; bk. 2; 2:84, 89–90; bk. 3.

64. Foxe 2:52–58, 64; bk. 3; 2:350–56; bk. 4.

65. Foxe 2:340–42; bk. 4.

66. Foxe 2:351; bk. 4.
67. Foxe 2:296–97; bk. 4.
68. Foxe 2:506; bk. 4.
69. Foxe 2:96; bk. 3.
70. Foxe 2:94–95; bk. 3.
71. Foxe 2:122; bk. 4.
72. Foxe 2:520–21; bk. 4.
73. Foxe 5:336–37; bk. 8.
74. Foxe 2:49, 98–99; bk. 3; 2:335; bk. 4; 5:288–90; bk. 8.
75. Foxe 2:266; bk. 4.
76. Foxe 2:117–18; bk. 4.
77. Foxe 2:264–70; bk. 4.
78. Foxe 2:356; bk. 4.
79. Foxe 2:196–52; bk. 4:

> If the cause make a martyr, as is said, I see not why we should esteem Thomas Becket to die a martyr, more than any others whom the prince's sword doth here temporally punish for their temporal deserts. . . . I suppose Thomas Becket to be far from the cause and title of a martyr, neither can he be excused from the charge of being a plain rebel against his prince.

80. Foxe 4:198–204; bk. 7; 4:643–51, 679, 688; bk. 8; 5:99–100; bk. 8.

81. Foxe 5:99–100; bk. 8. I have quoted only the last of More's jokes upon the scaffold.

82. Foxe 3:757; 4:135–36; bk. 6.

83. Foxe 4:249–50; bk. 7.

84. Foxe 2:777–78; bk. 5.

85. Foxe 5:136; bk. 8. The whole account runs from 134–36.

86. Foxe 5:46; bk. 8.

87. Foxe 5:260–61; bk. 8. See also 5:362–403; bk. 8, for Foxe's praise of Cromwell, whom he makes one of the great heroes of the faith, almost coequal with Wyclif, Hus, Tyndale, Luther, and Melanchthon: "a valiant soldier and captain of Christ, the foresaid lord Cromwell, as he was most studious of himself in a flagrant zeal to set forward the truth of the gospel, seeking all means and ways to beat down false religion and to advance the true." Notice the oppositional nature of Foxe's rhetoric: the advancement of true religion cannot be stated without a concomitant beating down of the false.

88. Foxe 2:724; bk. 5.

89. Foxe 2:726; bk. 5.

90. Foxe 7:574; bk. 11.

91. Foxe 6:658; bk. 11. See also the account of Ridley's pathetic cries of "I cannot burn!" at 7:550–51; bk. 11.

92. Foxe 5:438; bk. 8.

93. Cesare Baronio, *Martyrologium Romanum* . . . (Venice, 1584).

94. Theodore E. Mommsen, "Petrarch's Conception of the 'Dark Ages,'" *Speculum: A Journal of Medieval Studies* 17 (1942): 226–42.

95. See Lucie Varga, *Das Schlagwort vom "finsteren Mittelalter"* (Baden: Rohrer, 1932) 5; Mommsen 227.

96. Francesco Petrarca, *De sui ipsius et multorum ignorantia*, ed. L. M. Capelli (Paris: Champion, 1906) 45: ". . . paucis enim ante Cristi ortum obierat oculosque clauserat, heu! quibus e proximo noctis erratice ac tenebrarum finis et veritatis initium, vereque lucis aurora et iustitie sol instabat" (Mommsen 227).

97. Mommsen 231.

98. Mommsen 234.

99. Mommsen 234–37.

100. Mommsen 237.

101. Mommsen 240.

102. Flavio Biondo, *Historiarum ab inclinatione Romanorum imperii decades* . . . (Venice: 1483). See Denys Hay, *Flavio Biondo and the Middle Ages: Italian Lecture, British Academy, 1959*, Proceedings of the British Academy 45:98–127 (London: Oxford University Press, 1959), and Donald J. Wilcox, *In Search of God and Self: Renaissance and Reformation Thought* (Boston: Houghton, 1975) 55.

103. Wilcox, *In Search of God and Self* 54–56. See also his *The Development of Florentine Humanist Historiography in the Fifteenth Century*, Harvard Historical Studies 82 (Cambridge, Mass.: Harvard University Press, 1969).

104. Giorgio Vasari, *Vite degli artefici*, in *Opere di Giorgio Vasari, pittore*, Biblioteca enciclopedica Italiana 2 (Milan, 1829) 38, 62; *Lives of the Artists*, trans. George Bull (Harmondsworth: Penguin, 1965) 34, 49.

105. See Werner Goetz, *Translatio imperii; ein Beitrag zur Geschichte des Geschichtsdenkens und der politischen Theorien in Mittelalter und in der frühen Neuzeit* (Tübingen: Mohr, 1958). See also W. K. Ferguson, *Europe in Transition: 1300–1520* (Boston: Houghton, 1962).

106. Mommsen 234–36. There had been detractors of the *translatio imperii* in the Middle Ages (see Ernst Breisach, *Historiography: Ancient, Medieval, and Modern* [Chicago: University of Chicago Press, 1983] 163–65), but these had been of the minority and were of the losing faction.

107. See C. C. Bayley, "Petrarch, Charles IV, and the 'renovatio imperii,'" *Speculum: A Journal of Medieval Studies* 17 (1942): 323–41.

108. See E. B. Fryde, *Humanism and Renaissance Historiography* (London: Hambledon, 1983) 13–14.

109. See particularly Charles L. Stinger, *Humanism and the Church Fathers: Ambrogio Traversari (1386–1439) and Christian Antiquity in the Italian Renaissance* (Albany: State University of New York Press, 1977).

110. Luther, *Responsio ad condemnationem doctrinalem per Lovan. et Colon. factam*, Weimar 6:183. See also Headley 235.

111. See Herschel Baker, *The Race of Time: Three Lectures on Renaissance Historiography* (Toronto: University of Toronto Press, 1967) 34–41, and Arthur B. Ferguson, *Clio Unbound: Perception of the Social and Cultural Past in Renaissance England*, Duke Monographs in Medieval and Renaissance Studies 2 (Durham, N.C.: Duke University Press, 1979) 76.

112. Baker 60–62.

113. Lorenzo Valla, *The Treatise of Lorenzo Valla on the Donation of Constantine: Text and Translation into English*, ed. and trans. Christopher B. Coleman (New Haven: Yale University Press; London: Oxford University Press, 1922) 94–95.

114. See Fryde 12–13.

115. For a description of the new consciousness of time and its relations particularly to the development of mechanical timekeeping, see Ricardo J. Quinones, *The Renaissance Discovery of Time*, Harvard Studies in Comparative Literature 31 (Cambridge, Mass.: Harvard University Press, 1972).

116. See Wilcox, *In Search of God and Self* 62–65.

117. For the effects of print-culture, see Elizabeth L. Eisenstein, *The Printing Press as an Agent of Change: Communications and Cultural Transformations in Early-Modern Europe*, 2 vols. (Cambridge: Cambridge University Press, 1979).

118. Foxe 3:718–19; bk. 6.

119. Foxe 2:189, 271, 277, 534; bk. 4; 6:543; bk. 10. See also Wilcox, *The Development of Florentine Humanist Historiography* 102–5, for similar attitudes in Bruni.

120. Francis Bacon, *The Advancement of Learning*, in *The Works of Francis Bacon*, ed. J. Shedding, R. L. Ellis, and D. D. Heath, 14 vols. (London, 1857–74) 4:118. Cited by Ferguson, *Clio Unbound* 129.

Chapter 4

1. Edward Johnson, *Wonder-Working Providence of Sions Saviour in New England* (first published London, 1654), ed. J. Franklin Jameson, Original Narratives of Early American History (New York: Scribner's, 1910) 25–26; bk. 1, chap. 1.

2. William Bradford, *History of Plymouth Plantation*, ed. Charles Dean, Collections of the Massachusetts Historical Society 4th ser. 3 (Boston, 1856). This is the first publication of the manuscript left by Bradford at his death in 1657.

3. Bradford 1; bk. 1, chap. 1.

4. Bradford 1–3; bk. 1, chap. 1.

5. Bradford 3; bk. 1, chap. 1. Andrew Delbanco has discussed the importance of self-imposed intellectual isolation to the Puritan identity. See "The Puritan Errand Re-Viewed," *Journal of American Studies* 18 (1984): 343–60, and *The Puritan Ordeal* (Cambridge, Mass.: Harvard University Press, 1989).

6. Bradford 3–4; bk. 1, chap. 1.

7. Bradford 5; bk. 1, chap. 1.

8. See Jesper Rosenmeier, "'With my owne eyes': William Bradford's *Of Plymouth Plantation*," in *The American Puritan Imagination: Essays in Revaluation*, ed. Sacvan Bercovitch (London: Cambridge University Press, 1974) 77–106. The essay is also printed in *Typology and Early American Literature*, ed. Sacvan Bercovitch ([Amherst]: University of Massachusetts Press, 1972) 69–105. Rosenmeier takes as the starting point of his argument Bradford's abandonment of his history for the private study of Hebrew, which Rosenmeier interprets as a longing for an ahistorical knowledge of the past, "to be present in the past . . . [an] effort to resurrect the literal language of some original perfection."

9. Bradford 9; bk. 1, chap. 1.

10. Bradford 8–9; bk. 1, chap. 1.

11. Bradford 9–10; bk. 1, chap. 1.

12. Bradford 16–17; bk. 1, chap. 3.

13. Bradford 24; bk. 1, chap. 4.

14. Bradford 19; bk. 1, chap. 3.

15. Sacvan Bercovitch, *The Puritan Origins of the American Self* (New Haven: Yale University Press, 1975) 44–46: "A Separatist on the saint's course to heaven, Bradford assumes the traditional dichotomy between secular and sacred, and he sees the plantation itself, accordingly, in terms of common providence. To be sure, with all Protestants of his time he rejoices in the progress of the church. . . . But as a Separatist he expects no more from his own congregation than that it should hold fast to the principles of spiritual Israel; and as historian (not church historian) of Plymouth, he chronicles the fate of a wholly temporal venture."

16. Bradford 3; bk. 1, chap. 1.

17. Bradford 9; bk. 1, chap. 1.

18. This pun I borrow from Robert Lowell, "Children of Light," *Lord Weary's Castle* (1944; New York: Harvest-Harcourt, 1974) 34: "Pilgrims unhouseled by Geneva's Night, / They planted here the serpent's seeds of light." This archaic-sounding version of unhoused is actually from *housel*, the Middle and Modern English form of Saxon *husel* or *husl*, the Eucharist or Communion. To be unhouseled is both to be cast out from Communion and to be cast out from communion, unhoused.

19. Bradford 88; bk. 1, chap. 10. Also see Cecilia Tichi, "The Puritan Historians and Their New Jerusalem," *Early American Literature* 6 (1971): 143–

55. Tichi details the use of metaphors of building in the Puritan New England historians ("For the early decades of New England's growth the historians . . . favor metaphors that describe durable structures erected by cooperative effort") and also the ways in which these metaphors (temple, stones, foundation, studs, pillars) tend to disintegration in times of perceived declension.

20. Alan Howard sees the history as a "downward curve of failing strength," followed by "the ascent which measures the strength of God's sustaining hand." John Griffith interprets it as "a mercantile epic" of success. David Levin, by contrast, finds an unresolved oscillation between disease and remedy, adversity and prosperity, a complex equilibrium both of growth from small beginnings and of inevitable decline from original purity. Walter Wenska argues, with reference to the chronology of composition, that the two books should be regarded as two different histories, the first a story of establishment, the second one of declension. Kenneth Hovey outlines a series of cycles of divine wrath followed by repentance and deliverance. See Alan Howard, "Art and History in Bradford's *Of Plymouth Plantation*," *William and Mary Quarterly* 3rd ser. 28 (1971): 237–66; John Griffith, "*Of Plymouth Plantation* as a Mercantile Epic," *Arizona Quarterly* 28 (1972): 231–42; David Levin, "William Bradford: The Value of Puritan Historiography," in *Major Writers of Early American Literature*, ed. Everett Emerson (Madison: University of Wisconsin Press, 1972) 11–31; Walter P. Wenska, "Bradford's Two Histories: Pattern and Paradigm in *Of Plymouth Plantation*," *Early American Literature* 13 (1978): 151–64; Kenneth Alan Hovey, "The Theology of History in *Of Plymouth Plantation* and Its Predecessors," *Early American Literature* 10 (1975): 47–66. See also Robert Daly, "William Bradford's Vision of History," *American Literature* 44 (1973): 557–69.

21. Bradford 89; bk. 2.

22. See Wenska 151–64.

23. The following are a few examples of the epithets applied to the land by Puritan writers: "Desart Wildernesse," Edward Johnson 43; bk. 1, chap. 8; "this dismal Dessart," Johnson 48; bk. 1; chap. 10; "You have solemnly professed before God, angels, and men that the cause of your leaving your country, kindred, and fathers' houses and transporting yourselves with your wives, little ones, and substance over the vast ocean into this waste and howling wilderness. . . ," Samuel Danforth, *A Brief Recognition of New Englands Errand into the Wilderness; Made in the Audience of the General Assembly of the Massachusets Colony, at Boston in N. E.* (Cambridge, Mass., 1671), rpt. in *The Wall and the Garden: Selected Massachusetts Election Sermons 1670–1775*, ed. A. W. Plumstead (Minneapolis: University of Minnesota Press, 1968) 65; "poor men that were now to transplant themselves into an horrid *wildernesse*," Cotton Mather, *Magnalia Christi Americana: Or, the Ecclesiastical History of New-England, from its First Planting in the Year 1620, unto the*

Year of Our Lord, 1698, 1st American ed., 2 vols. ([London, 1702]; Hartford, 1820) 1:47; bk. 1, chap. 2; "this land, which was then an hideous howling wilderness," Jonathan Edwards, *A History of the Work of Redemption, Containing the Outlines of a Body of Divinity, in a Method Entirely New,* 2nd ed. (Edinburgh, 1799) 334; part 2, sec. 1, period 3.

24. Samuel Sewall, *Phaenomena quaedam Apocalyptica ad Aspectum Novi Orbis Configurata: Or, Some Few Lines towards a Description of the New Heaven as It Makes to Those Who Stand upon the New Earth* (Boston, 1697) 59. See Perry Miller and Thomas H. Johnson, eds., *The Puritans,* rev. ed., 2 vols. (New York: Harper, 1963) 1:290.

25. Bradford 78–79; bk. 1, chap. 9.

26. Bercovitch, *The Puritan Origins of the American Self* 44–46.

27. Bradford 382–83; bk. 2.

28. Bradford 384; bk. 2.

29. Bradford 384–85; bk. 2.

30. Bradford 385–86; bk. 2.

31. Bradford 397–98; bk. 2.

32. Bradford 427; bk. 2.

33. Bradford 444; bk. 2. For comment on these entries see Bercovitch, *The Puritan Origins of the American Self* 46; Wenska 157.

34. Rosenmeier 77–78.

35. See Ursula Brumm, "Edward Johnson's *Wonder-Working Providence* and the Puritan Conception of History," *Jahrbuch für Amerikastudien* 14 (1969): 140–51; and Edward J. Gallagher, "An Overview of Edward Johnson's *Wonder-Working Providence,*" *Early American Literature* 5 (1971): 30–49.

36. Brumm 141.

37. Johnson 23; bk. 1, chap. 1.

38. The major studies are Sacvan Bercovitch, ed., *Typology and Early American Literature* ([Amherst]: University of Massachusetts Press, 1972); Bercovitch, *The American Jeremiad* (Madison: University of Wisconsin Press, 1978), particularly chap. 4; Ursula Brumm, *Die religiöse Typologie im amerikanischen Denken; ihre Bedeutung für die amerikanische Literatur- und Geistesgeschichte,* Studien zur amerikanische Literatur und Geschichte 2 (Leiden: Brill, 1963); Mason I. Lowance, *The Language of Canaan: Metaphor and Symbol in New England from the Puritans to the Transcendentalists* (Cambridge, Mass.: Harvard University Press, 1980). For a specifically typological study of Johnson, see Bercovitch, "The Historiography of Johnson's *Wonder-Working Providence,*" *Essex Institute Historical Collections* 104 (1968): 138–61.

39. Augustine, for one, found much in the Old Testament absurd and offensive until he heard Ambrose expound it typologically: *Confessions* bk. 6, chaps. 3–5.

40. Mason Lowance makes this point in "Typology and the New England Way: Cotton Mather and the Exegesis of Biblical Types," *Early American Literature* 4 (1969): 15–37.

41. John Winthrop, *The History of New England from 1630 to 1649* [Journal], ed. James Savage, new ed., 2 vols. (Boston, 1853) 2:24.

42. Two examples of this Puritan obsession: "[The Song of Songs is] a divine abridgement of the Acts and Monuments of the Church [and is] historical Prophesie or Prophetical history," John Cotton, *A Briefe Exposition of the Whole Book of Canticles, or, Song of Solomon* . . . (London, 1648), cited in Brumm, "Edward Johnson's *Wonder-Working Providence* and the Puritan Conception of History" 141; "*Prophesie* is *Historie antedated*; and *Historie* is *Postdated Prophesie*," Nicholas Noyes, *New-Englands Duty* (Boston, 1698), cited in Bercovitch, *The American Jeremiad* 15.

43. Johnson 24; bk. 1, chap. 1.

44. Johnson 24–25; bk. 1, chap. 1.

45. Rev. 19:15.

46. Rev. 5:5–6.

47. Johnson 25–26; bk. 1, chap. 2.

48. Johnson 87; bk. 1, chap. 27.

49. Johnson 147; bk. 2, chap. 1.

50. Johnson 168; bk. 2, chap. 6.

51. Bradford 22; bk. 1, chap. 4.

52. Sacvan Bercovitch has framed part of the Puritan legacy in terms of a "ritual of consensus": *The American Jeremiad* 131–75.

53. Johnson 27; bk. 1, chap. 3; 28–30; bk. 1, chap. 4.

54. Johnson 32; bk. 1, chap. 5.

55. Johnson 31; bk. 1, chap. 5; 50; bk. 1, chap. 11.

56. Johnson 36; bk. 1, chap. 6.

57. Johnson 49; bk. 1, chap. 11.

58. Johnson 40; bk. 1, chap. 8.

59. See Johnson 41–43; bk. 1, chap. 8; 147–50; bk. 2, chap. 1; 168–70; bk. 2, chap. 6; Bradford 357; bk. 2. See also Increase Mather, *A Relation of the Troubles which Have Hapned in New-England, by Reason of the Indians There* . . . (Boston, 1677). Bradford is the only chronicler of the destruction of the Pequods who expresses sympathy with the Indians and horror at their slaughter, while still approving the colonists' actions. Increase Mather reveals throughout a fascinating conjunction of antipathy for the Indians (expressed by concentrated use of destruction-of-peoples typology) with genuine concern for their salvation. Two modern studies of the subject are Alden T. Vaughan, *New England Frontier: Puritans and Indians, 1620–1675* (Boston: Little, 1965), and Francis Jennings, *The Invasion of America: Indians, Colonialism and the Cant of Conquest* (Chapel Hill: University of North Carolina Press, 1975), the latter an extended exercise in anti-Puritan polemic.

60. Johnson 49; bk. 1, chap. 11.

61. Johnson 25; bk. 1, chap. 1.

62. Johnson 34–35; bk. 1, chap. 6.

63. Johnson 122; bk. 1, chap. 39.

64. The immediate origins of this earthly Second Advent can be found in Thomas Brightman, *Apocalypsis Apocalypseos, or a Revelation of the Revelation . . .* (Leiden, 1616). For Brightman's place in Reformation historiography, see Avihu Zakai, "Reformation, History, and Eschatology in English Protestantism," *History and Theory: Studies in the Philosophy of History* 26 (1987): 300–318. See also Richard Bauckham, *Tudor Apocalypse: Sixteenth Century Apocalypticism, Millenarianism and the English Reformation: From John Bale to John Foxe and Thomas Brightman* (Oxford: Sutton Courtney, [1978?]), and Cecilia Tichi, *New World, New Earth: Environmental Reform in American Literature from the Puritans through Whitman* (New Haven: Yale University Press, 1979) 19. Ernest Lee Tuveson has explored connections between earthly chiliasm, elect nation theory, and secularization in *Redeemer Nation: The Idea of America's Millennial Role* (Chicago: University of Chicago Press, 1968) and *Millennium and Utopia: A Study in the Background of the Idea of Progress* (New York: Harper, 1964).

65. Johnson 52; bk. 1, chap. 12.

66. Johnson 53; bk. 1, chap. 12.

67. Johnson 59; bk. 1, chap. 15.

68. See, for example, 124–36; bk. 1, chaps. 40–43.

69. Johnson 252–57; bk. 2, chap. 8.

70. Joshua Scottow, *A Narrative of the Planting of the Massachusets Colony Anno 1628. With the Lords Signal Presence the First Thirty Years. Also a Caution from New-Englands Apostle, the Great Cotton, How to Escape the Calamity, which Might Befall Them or Their Posterity. And Confirmed by the Evangelist Norton with Prognosticks from the Famous Dr. Owen. concerning the Fate of these Churches, and Animadversions upon the Anger of God, in Sending of Evil Angels among Us. Published by Old Planters, the Authors of the Old Mens Tears,* Collections of the Massachusetts Historical Society 4th ser. 4 (Boston, 1858) 279–330; first published Boston, 1694.

71. Scottow (ca. 1618–1698) was of the first generation; he had come to Massachusetts before 1634, and, although younger than Edward Johnson (1598–1672), his years of experience in the colony are almost identical to those of Johnson up to Johnson's death. Johnson came to Massachusetts in 1630 with Winthrop, but was absent in England from 1631–1636. See J. Franklin Jameson, Introduction, *Wonder-Working Providence* 3–18.

72. Scottow 283.

73. For an analysis of Scottow's use of figuration and its subsumption of the "narrative" of the title, see Dennis Powers, "Purpose and Design in Joshua Scottow's *Narrative,*" *Early American Literature* 18 (1984): 275–90.

74. Scottow 287.

75. Scottow 284–85.

76. Herman Melville, *Moby-Dick* chap. 8.

77. See Scottow 327: "*It concerneth* New-England *always to remember, that originally they are a* PLANTATION *Religious, not a* PLANTATION *of Trade.*"

78. Scottow 288.

79. See Scottow 287, 298–99: "thus the Motions of the Wheels were very high and terrible to our Adversaries, and so were the Rings and Wheels of Providence full of Eyes."

80. Scottow 310–11.

81. Scottow 293, 301.

82. For the concept and mechanism of centrifugal Reformation, see William Haller, *The Rise of Puritanism: Or, the Way to the New Jerusalem as Set Forth in Pulpit and Press from Thomas Cartwright to John Lilburn and John Milton, 1570–1643* (New York: Columbia University Press, 1938) 16.

83. It is significant that the intellectual rebellion that arose against the ministers after the witch trials of 1692–1693 was led by such merchants as Robert Calef and Thomas Brattle, as the merchant classes always have great interest in the physical coherence of a society. See David Levin, *Cotton Mather: The Young Life of the Lord's Remembrancer, 1663–1703* (Cambridge, Mass.: Harvard University Press, 1978) 47, 82, 120, 182, 221, 223, 240–45, 247–48, 274, 286–88, 290–93, 295–96.

84. Scottow 301.

85. Scottow 327.

86. Scottow 326.

87. Cotton Mather 1:23; General Introduction.

88. Sacvan Bercovitch has called attention the self-conscious, epic literariness of the *Magnalia* in "New England Epic: Cotton Mather's *Magnalia Christi Americana,*" *ELH* 33 (1966): 337–50.

89. Mather 1:24–25; General Introduction.

90. Mather 1:25; General Introduction.

91. Mather 1:24; General Introduction.

92. Mather 2:183; bk. 5, chap. 1.

93. Mather 1:41; bk. 1, chap. 1.

94. Mather 1:40; bk. 1, Introduction.

95. Mather 1:25–26; General Introduction.

96. Mather 1:39; bk. 1, Introduction.

97. See Lowance, *The Language of Canaan* 154, for Mather's abandonment of the earthly and American millennium.

98. Mather 1:26; General Introduction.

99. See for instance 1:26–27; General Introduction, in which he cites all of

the classical historians as models of history, only to dismiss them as incompatible with Christianity and the simplicity of the Gospels and Acts.

100. Mather 1:79; bk. 1, chap. 7.

101. For relations between biography, troubled times, and the disintegration of the communal, see Kenneth B. Murdock, "Clio in the Wilderness: History and Biography in Puritan New England," *Church History* 24 (1955): 221–38; also reprinted in *Early American Literature* 6 (1972): 201–19.

102. Mather 2:591, bk. 7.

103. Edwards 342; pt. 2, sec. 1.

104. Edwards 343–44; pt. 2, sec. 1.

105. Peter Gay finds the sense of intellectual isolation in Edward's history so acute that he subtitles his chapter on it "An American Tragedy." See Peter Gay, *A Loss of Mastery: Puritan Historians in Colonial America* (Berkeley: University of California Press, 1966) 88–117.

106. Edwards 345; pt. 1, sec. 1.

107. Edwards 337; bk. 1, sec. 1.

108. Benjamin Franklin, *Autobiography*, in *The Works of Benjamin Franklin*, ed. John Bigelow, 12 vols. (New York: Knickerbocker-Putnam's, 1904) 1:224–25: "Being among the hindmost in Market-street, I had the curiosity to learn how far he could be heard, by retiring backwards down the street towards the river; and I found his voice distinct till I came near Front-street when some noise in that street obscur'd it. Imagining then a semicircle, of which my distance should be the radius and that it were fill'd with auditors, to each of whom I allow'd two square feet, I computed that he might well be heard by more than thirty thousand. This reconcil'd me to the newspaper accounts of his having preach'd to tweny-five thousand people in the fields and to the antient histories of generals haranguing whole armies, of which I had some times doubted."

Epilogue

1. Ralph Waldo Emerson, *Nature*, in *The Collected Works of Ralph Waldo Emerson*, ed. Robert E. Spiller and Alfred R. Ferguson, 4 vols. (Cambridge, Mass.: Belknap Press, Harvard University Press, 1971–1987) 1:7; Introduction.

2. Emerson, "The American Scholar: An Oration Delivered Before the Phi Beta Kappa Society, at Cambridge, August 31, 1837," in *Collected Works* 1:52, 56.

3. Franklin's statement is typical: ". . . every other sect supposing itself in possession of all truth, and that those who differ are so far in the wrong; like a man travelling in foggy weather, those at some distance before him on the road

he sees wrapped up in the fog, as well as those behind him, and also the people in the fields on each side, but near him all appears clear, tho' in truth he is as much in the fog as any of them." *Autobiography*, in *The Works of Benjamin Franklin*, ed. John Bigelow, 12 vols. (New York: Knickerbocker-Putnam's, 1904) 1:237.

4. The Unitarian revolution has its doctrinal, or antidoctrinal, origins in some of the earliest Anabaptist groups, and the revolution culminated in the founding of the American Unitarian Association in Massachusetts in 1825. For the rationalist revolution within the Church, see Gerald Robertson Cragg, *From Puritanism to the Age of Reason: A Study of Changes in Religious Thought within the Church of England, 1660–1700* (Cambridge: Cambridge University Press, 1950); Peter Gay, *The Enlightenment, An Interpretation: The Rise of Modern Paganism* (New York: Knopf, 1966); Gustav Adolf Koch, *Republican Religion: The American Revolution and the Cult of Reason* (New York: Holt, 1933), republished as *Religion of the American Enlightenment* (New York: Crowell, 1968); Frank Edward Manuel, *The Eighteenth Century Confronts the Gods* (Cambridge, Mass.: Harvard University Press, 1959); Roland N. Stromberg, *Religious Liberalism in Eighteenth-Century England* (London: Oxford University Press, 1954).

5. Henry David Thoreau, *Walden; or, Life in the Woods*, in *Henry David Thoreau: A Week on the Concord and Merrimack Rivers, Walden; or, Life in the Woods, The Maine Woods, Cape Cod*, ed. Robert F. Sayre (New York: Library of America, 1985) 329, 331; chap. 1.

6. Emerson, *Nature* 1:7; Introduction.

7. Thoreau exhibits the same paradigm in the third chapter of *Walden*, "Reading," in which he asserts that the previous generations could not understand the classics because of the depravity and unnaturalness of their culture.

8. Arthur Hugh Clough, "Paper on Religion," in *Selected Prose Works of Arthur Hugh Clough*, ed. Buckner B. Trawick (University: University of Alabama Press, 1964) 287–88.

9. Quoted with ironic approbation by Henry Adams, *The Education of Henry Adams*, in *Henry Adams: Novels, Mont Saint Michel, The Education*, ed. Ernest Samuels and Jane N. Samuels (New York: Library of America, 1983) 1073; chap. 25.

10. Friedrich Nietzsche, *Von Nutzen und Nachteil der Historie für das Leben* (1874), in *Friedrich Nietzsche: Sämtliche Werke*, ed. Giorgio Colli and Mazzino Montinari, 15 vols. (Berlin: Walter de Gruyter, 1967–1977) 1:329; *On the Advantage and Disadvantage of History for Life*, trans. Peter Preuss (Indianapolis: Hackett, 1980) 62.

11. Nietzsche, *Friedrich Nietzsche: Sämtliche Werke* 1:271–72; *On the Advantage and Disadvantage of History for Life* 23.

12. Nietzsche, *Friedrich Nietzsche: Sämtliche Werke* 1:295–96; *On the Advantage and Disadvantage of History for Life* 38–39.

13. Nietzsche, *Friedrich Nietzsche: Sämtliche Werke* 1:329; *On the Advantage and Disadvantage of History for Life* 62.

14. Nietzsche, *Friedrich Nietzsche: Sämtliche Werke* 1:308; *On the Advantage and Disadvantage of History for Life* 47.

15. George Bancroft, *History of the United States of America, from the Discovery of the Continent*, 6 vols. (New York, 1883–1885) 1:3; Introduction.

16. Thomas Hobbes, *Leviathan: Or the Matter, Forme and Power of a Commonwealth Ecclesiastical and Civil*, ed. Michael Oakeshott (New York: Collier-Macmillan, 1962); John Locke, *An Essay concerning Human Understanding*, ed. P. Nidditch (Oxford: Clarendon, 1975); Giambattista Vico, *Principi di scienza nuova di Giambattista Vico d'intorno alla comune natura delle nazioni*, 3rd ed. (Naples, 1744); *The New Science of Giambattista Vico: Unabridged Translation of the Third Edition (1744) with the Addition of "Practic of the New Science,"* trans. Thomas Goddard Bergin and Max Harold Fisch (Ithaca, N.Y.: Cornell University Press, 1984).

17. London, 1859.

18. Joseph Arthur, Comte de Gobineau, *Essai sur l'inégalité des races humaines*, 4 vols. (Paris, 1853–1855); *The Inequality of Human Races*, trans. Adrian Collins (London: Heinemann, 1915); Houston Stewart Chamberlain, *Rasse und Persönlichkeit* (Munich: Bruckmann, 1925).

19. For a general account of developmental history, see Peter J. Bowler, *Evolution: The History of an Idea* (Berkeley: University of California Press, 1984). For more specific treatment of the connections between process and race, see Harold E. Pagliaro, ed., *Racism in the Eighteenth Century*, Studies in Eighteenth-Century Literature 3 (Cleveland: Case Western Reserve University Press, 1973); John S. Haller, *Outcasts from Evolution: Scientific Attitudes of Racial Inferiority, 1859–1900* (Urbana: University of Illinois Press, 1975); Louis L. Snyder, *The Idea of Racialism: Its Meaning and History* (Princeton: Princeton University Press, 1962); Ashley Montagu, *Race, Society and Humanity* (New York: Van Nostrand Reinhold, 1963), and *Man's Most Dangerous Myth: The Fallacy of Race*, 5th ed. (Oxford: Oxford University Press, 1974); Jacques Barzun, *Race: A Study in Superstition* (New York: Harcourt, 1965).

20. Nietzsche, *Friedrich Nietzsche: Sämtliche Werke* 1:272; *On the Advantage and Disadvantage of History for Life* 24.

21. Nietzsche, *Friedrich Nietzsche: Sämtliche Werke* 1:327–29; *On the Advantage and Disadvantage of History for Life* 60–61.

22. Vico, *The New Science* 62–63; bk. 1, sec. 2.

23. Vico, *The New Science* 67; bk. 1, sec. 2, axiom 22.

24. Charles Louis de Secondat, baron de Brède et de Montesquieu, *De l'esprit des laix* . . . (Geneva, 1748).

25. Ernst Cassirer, *Die Philosophie der Aufklärung*, Grundriss der philosophischen Wissenschaften (Tübingen: Mohr, 1932) 266–67; *The Philosophy of*

the Enlightenment, trans. Fritz C. A. Koelin and James P. Pettegrove (Princeton: Princeton University Press, 1951) 199. The title of Cassirer's chapter is appropriately "The Conquest of the Historical World."

26. Hobbes 267–68; pt. 2, chap. 31.

27. Thomas Kuhn, *The Copernican Revolution: Planetary Astronomy in the Development of Western Thought* (Cambridge, Mass.: Harvard University Press, 1957); *The Structure of Scientific Revolutions* (Chicago: University of Chicago Press, 1959).

28. Joseph Juste Scaliger, *De emendatione temporum* . . . (Frankfurt, 1593); James Ussher, *Jacobi Usserii Armachani annales veteris et Novi Testamenti* . . . (Paris, 1673).

29. Ernst Breisach, *Historiography: Ancient, Medieval, and Modern* (Chicago: University of Chicago Press, 1983) 197.

30. David Ramsey, *The History of the American Revolution* (Philadelphia, 1789); Jedidiah Morse, *The History of America, in Two Books* (Philadelphia, [n.d.]); Noah Webster, *History of the United States; To Which Is Prefixed a Brief Historical Account of Our English Ancestors, from the Dispersion of Babel, to Their Migration to America; and of the Conquest of South America, by the Spaniards* (New Haven, 1832); Mason Locke Weems, *The Life of George Washington; with Curious Anecdotes, Equally Honorable to Himself, and Exemplary to His Young Countrymen*, 6th ed. (Philadelphia, 1808).

31. Bancroft 1:2; Introduction.

32. Bancroft 1:3.

33. Adams, *The Education* 1117; chap. 29.

34. Adams, *Mont Saint Michel and Chartres*, in *Henry Adams: Novels, Mont Saint Michel, The Education*, ed. Ernest Samuels and Jane N. Samuels (New York: Library of America, 1983) 343; chap. 1.

35. Adams, *Mont Saint Michel* 345; chap. 1.

36. Adams, *Mont Saint Michel* 347, 349; chap. 1.

37. Adams, *Mont Saint Michel* 371, 382–83; chap. 3.

38. Adams, *Mont Saint Michel* 439; chap. 7.

39. Adams, *Mont Saint Michel* 372–73; chap. 3.

40. Adams, *Mont Saint Michel* 409; chap. 5.

41. Adams, *Mont Saint Michel* 694–95; chap. 16.

42. Adams, *The Education* 722; Preface.

43. Adams, *The Education* 763, 765; chap. 3.

44. Adams, *The Education* 1112; chap. 29.

45. T. S. Eliot, "Gerontian."

46. Adams, *The Education* 863; chap. 10.

47. Adams, *The Education* 925; chap. 15.

48. Adams, *The Education* 776; chap. 4.

49. Adams, *The Education* 777; chap. 4. See Jarel C. Sallee, "Henry Adams'

Emersonian Education," *ESQ: A Journal of the American Renaissance* 27 (1981): 38–46.

50. Adams, *The Education* 928; chap. 15. See Paul J. Hamill, "Science as Ideology: The Case of the Amateur, Henry Adams," *Canadian Review of American Studies* 12 (1981): 21–35; and Howard N. Munford, "Henry Adams: The Limitations of Science," in *Critical Essays on Henry Adams*, ed. Earl N. Harbert (Boston: Hall, 1981).

51. Adams *The Education* 929, 931; chap. 15.

52. Karl Marx, "Speech at the Anniversary of the *People's Paper*," in *The Marx-Engels Reader*, ed. Robert C. Tucker, 2nd ed. (New York: Norton, 1978) 577; also cited by Marshall Berman, *All That Is Solid Melts into Air: The Experience of Modernity* (Harmondsworth: Penguin, 1988) 20.

53. Adams, *The Education* 982–84; chap. 19.

54. Adams, *The Education* 1020–21; chap. 21.

55. Adams, *The Education* 1068–69; chap. 25.

56. Adams, *The Education* 1091; chap. 27.

57. This cultural remnant of christianity (to preserve Adams' lower case) is mediated for him by Hegel and Schopenhauer, on which see Joseph G. Kronick, "The Limits of Contradiction: Irony and History in Hegel and Henry Adams," *CLIO: A Journal of Literature, History, and the Philosophy of History* 15 (1986): 391–410.

58. Adams, *The Education* 1138; chap. 31.

59. Adams, *The Education* 1138; chap. 31.

60. Adams, *The Education* 1173; chap. 34.

61. Adams, *The Education* 1138–39, 1140–41; chap. 31.

62. Adams, *The Education* 1066–76; chap. 25.

63. Adams, *The Education* 1110–11; chap. 29.

64. Adams, "Prayer to the Virgin of Chartres," in *Henry Adams: Novels, Mont Saint Michel, The Education*, ed. Ernest Samuels and Jane N. Samuels (New York: Library of America, 1983) 1202–7.

65. Adams, *The Education* 1112; chap. 29.

66. Adams, *The Education* chaps. 33 and 34; 1153–75. See Wayne Lesser, "Criticism, Literary History and the Paradigm: *The Education of Henry Adams*," *PMLA: Publications of the Modern Language Association of America* 97 (1982): 378–94: "[The dynamic theory of history is a] revisionist reading of our failed encounter with nature . . . our resolute refusal to accept our political, social, spiritual, and personal representations as conditional."

67. Adams, *The Education* 1132; chap. 31.

68. Adams, *The Education* 1138; chap. 31.

69. Adams, *The Education* 1151; chap. 32.

70. As correlative and proof I offer this sentence from a 1958 lecture by Robert Oppenheimer: "But it is very clear that if knowledge is being changed

every few years, arching structures like values, which unite the present and the future, which unite them—the past and the future through the present—are going to have great trouble retaining their content, retaining their factual, practical meaning." Robert Oppenheimer, "Knowledge and the Structure of Culture," Helen Kenyon Lecture, Vassar College, 29 October 1958, unpublished typescript, University of California, Los Angeles. I am indebted to Robert Faggen for bringing this piece to my notice.

71. Hayden White, *Metahistory: The Historical Imagination in Nineteenth-Century Europe* (Baltimore: Johns Hopkins University Press, 1973) xi.

72. White xii.

73. For an account of this see Hayden White, "The Burden of History," *History and Theory* 5 (1966): 111–34; Louis O. Mink, "Philosophical Analysis and Historical Understanding," *Review of Metaphysics* 21 (1968): 667–98; William H. Dray, ed., *Philosophical Analysis and History*, Sources in Contemporary Philosophy (New York: Harper, 1966).

74. Ludwig Wittgenstein, *Culture and Value*, ed. G. H. von Wright and Heikki Nyman, trans. Peter Winch (Chicago: University of Chicago Press, 1984) 85, 86, 85e, 86e.

75. For an example of skepticism outreaching skepticism, see Hayden White, "Foucault Decoded: Notes from Underground," *History and Theory: Studies in the Philosophy of History* 12 (1973): 23–34, in which White asserts "Foucault . . . does not seem to be aware that the categories he uses for analyzing the human sciences are little more than formalizations of the tropes," *Metahistory* 2, n. 4.

76. Simone Weil, "A propos de la question coloniale, dans ses rapports avec le peuple français," in *Écrits historiques et politiques* (Paris: Gallimard, 1960) 366; "East and West: Thoughts of the Colonial Problem," in *Selected Essays*, trans. Richard Rees (London: Oxford University Press, 1962) 197.

Bibliography

The following lists of sources make no claim to comprehensiveness. I have included only those studies that have a more or less direct bearing on the issues examined in this thesis. For more comprehensive and general bibliographies, see Breisach, Sterns, and Miller and Johnson below.

Primary Sources

Adams, Henry. *Henry Adams: Novels, Mont Saint Michel, The Education*. Ed. Ernest Samuels and Jane N. Samuels. New York: Library of America, 1983.

Alighieri, Dante. See Dante Alighieri.

Augustine of Hippo. *Saint Augustine: The City of God Against the Pagans*. Trans. varies. 7 vols. Loeb Classical Library. Cambridge, Mass.: Harvard University Press; London: Heinemann, 1957–1972.

————. *Saint Augustine's Confessions*. Trans. William Watts. 2 vols. Loeb Classical Library. Cambridge, Mass.: Harvard University Press; London: Heinemann, 1950–1951.

Bacon, Francis. *The Advancement of Learning*. Vol. 4 of *The Works of Francis Bacon*. Ed. James Spedding, R. L. Ellis, and D. D. Heath. 14 vols. London, 1858–1874.

Bancroft, George. *History of the United States of America, from the Discovery of the Continent*. 6 vols. New York, 1883–1885.

Baronio, Cesare. *Martyrologium Romanum*. . . . Venice, 1584.

Bede. *Historia ecclesiastica gentis Anglorum*. In *Baedae opera historica*, trans. J. E. King. 2 vols. Loeb Classical Library. London: Heinemann; New York: Putnam's, 1930.

Bernard of Clairvaux. *Opera Omnia*. Ed. D. Joannis Mabillon. 4th ed. 2 vols. Paris, 1839.

Bernard of Clairvaux. *The Letters of Saint Bernard of Clairvaux*. Trans. Bruno Scott James. Chicago: Henry Regnery, 1953.

Biondo, Flavio. *Historiarum ab inclinatione Romanorum imperii decades.* . . . Venice, 1483.

Bradford, William. *History of Plymouth Plantation*. Ed. Charles Dean. Collections of the Massachusetts Historical Society 4th ser. 3. Boston, 1856.

Brightman, Thomas. *Apocalypsis Apocalypseos, or a Revelation of the Revelation.* . . . Leiden, 1616.

Chamberlain, Houston Stewart. *Rasse und Persönlichkeit*. Munich: Bruckmann, 1925.

Chretien de Troyes. *Cligés*. Ed. Wendelin Foerster. Christian von Troyes sämtliche Werke. Halle, 1884.

————. *Chretien de Troyes: Arthurian Romances*. Trans. W. W. Comfort. Ed. D. D. R. Owen. Everyman's Library. London: Dent; New York: Dutton, 1914.

Clough, Arthur Hugh. "Paper on Religion." In *Selected Prose Works of Arthur Hugh Clough*. ed. Buckner B. Trawick. University: University of Alabama Press, 1964.

Cotton, John. *A Briefe Exposition of the Whole Book of Canticles, or, Song of Solomon.* . . . London, 1648.

Danforth, Samuel. *A Brief Recognition of New Englands Errand into the Wildernesse; Made in the Audience of the General Assembly of the Massachusets Colony, at Boston in N. E.* . . . Cambridge, Mass., 1671. Rpt. in *The Wall and the Garden: Selected Massachusetts Election Sermons 1670–1775*, ed. A. W. Plumstead. Minneapolis: University of Minnesota Press, 1968.

Dante Alighieri. *De monarchia*. Ed. Karl Witte. Vienna, 1874.

————. *The* De monarchia *of Dante Alighieri*. Trans. Aurelia Henry. Boston: Riverside-Houghton, 1904.

Darwin, Charles. *On the Origin of Species by Means of Natural Selection, Or the Preservation of Favoured Races in the Struggle for Life*. London, 1859.

Edwards, Jonathan. *A History of the Work of Redemption, Containing the Outlines of a Body of Divinity, in a Method Entirely New*. 2nd ed. Edinburgh, 1799.

Emerson, Ralph Waldo. *The Collected Works of Ralph Waldo Emerson*. Ed. Robert E. Spiller and Alfred R. Ferguson. 4 vols. Cambridge, Mass.: Belknap Press, Harvard University Press, 1971–1987.

England, George, and Alfred W. Pollard, eds. *The Townely Plays*. Early English Text Society, Extra Series 71. 1897. London: Oxford University Press, 1966.

Eusebius of Caesarea. *Eusebius: The Ecclesiastical History*. Trans. Kirsop

Lake. 2 vols. Loeb Classical Library. London: Heinemann; New York: Putnam's, 1926.

Foxe, John. *The Acts and Monuments of John Foxe.* Ed. Josiah Pratt. 8 vols. 1870. New York: AMS, 1965.

————. *Commentarii rerum in Ecclesia gestarum.* . . . Strasbourg, 1554.

Franklin, Benjamin. *Autobiography.* Vol. 1 of *The Works of Benjamin Franklin.* Ed. John Bigelow. 12 vols. New York: Knickerbocker-Putnam's, 1904.

Fulcher of Chartres. *Fulcheri Carnotensis historia Hierosolymitana.* Ed. Heinrich Hagenmeyer. Heidelberg: Winters, 1913.

————. *Fulcher of Chartres: Chronicle of the First Crusade.* Trans. Martha Evelyn McGinty. Translations and Reprints from the Original Sources of History 3rd ser. 1. Philadelphia: University of Pennsylvania Press; London: Oxford University Press, 1941.

Geoffrey of Monmouth. *The* Historia Regum Brittanie *of Geoffrey of Monmouth.* Ed. Acton Griscom. London: Longmans, 1929.

————. *History of the Kings of Britain.* Trans. Sebastian Evans. Rev. Charles W. Dunn. New York: Dutton, 1958.

Gobineau, Joseph Arthur, Comte de. *Essai sur l'inégalité des races humaines.* 4 vols. Paris, 1853–1855.

————. *The Inequality of Human Races.* Trans. Adrian Collins. London: Heinemann, 1915.

Gregory of Tours. *Gregoire de Tours: Histoire des Francs.* Ed. René Poupardin. Collection de textes pour servir à l'étude et à l'enseignement de l'histoire. Paris: Alphonse Picard, 1913.

————. *The History of the Franks.* Trans. Lewis Thorpe. Penguin Classics. Harmondsworth: Penguin, 1974.

Helmold. *The Chronicle of the Slavs.* Ed. Francis Joseph Tschan. Records of Civilization: Sources and Studies. New York: Columbia University Press, 1935.

Henry VIII. *Assertio septem sacramentorum adversus Martin Lutherus.* [London and Rome], 1521.

————. *Assertio septem sacramentorum: or, An Assertion of the Seven Sacraments against Martin Luther.* Trans. T[homas]. W[ebster]. London, 1687.

Hobbes, Thomas. *Leviathan: Or the Matter, Forme and Power of a Commonwealth Ecclesiastical and Civil.* Ed. Michael Oakeshott. New York: Collier-Macmillan, 1962.

Hooker, Thomas. *A Survey of the Summe of Church-Discipline.* . . . London, 1648.

Hus, Jan. *Ioannes Huss de anatomia antichristi.* . . . Strasbourg, 1524.

Johnson, Edward. *Wonder-Working Providence of Sions Saviour in New*

England. Ed. J. Franklin Jameson. Original Narratives of Early American History. New York: Scribner's, 1910.

Locke, John. *An Essay concerning Human Understanding.* Ed. P. Nidditch. Oxford: Clarendon, 1975.

Luther, Martin. *D. Martin Luthers Werke.* 61 vols. Weimar: Hermann Böhlau, 1883–1983.

————. *D. Martin Luthers Werke: Briefwechsel.* 18 vols. Weimar: Hermann Böhlau, 1930–1985.

————. *Luther's Works.* General eds. Jaroslav Pelikan and Helmut T. Lehman. 55 vols. Saint Louis: Concordia, 1958–1968.

Marx, Karl. "Speech at the Anniversary of the *People's Paper.*" In *The Marx-Engels Reader,* ed. Robert C. Tucker. 2nd ed. New York: Norton, 1978.

Mather, Cotton. *Magnalia Christi Americana, Or, the Ecclesiastical History of New-England, from its First Planting in the Year 1620, unto the Year of Our Lord, 1698.* 1st American ed. 2 vols. Hartford, 1820.

Mather, Increase. *A Relation of the Troubles which have Hapned in New-England, by Reason of the Indians There. . . .* Boston, 1677.

Matthew of Westminster. *Flores historiarum.* Ed. Henry Richards Luard. 2 vols. Chronicles and Memorials of Great Britain and Ireland during the Middle Ages (Rolls Series) 95. London, 1890.

————. *The Flowers of History.* Trans. C. D. Yonge. 3 vols. London, 1853.

Matthew Paris. See Paris, Matthew.

Monk of Saint Gall, The. *Monachus Sangallensis.* In *Monumenta Carolini,* ed. Philippe Jaffe. Bibliotheca rerum Germanicarum. Berlin, 1867.

————. *Early Lives of Charlemagne, by Eginhard and the Monk of Saint Gall.* Trans. and ed. A. J. Grant. London: Chatto, 1922.

Montesquieu, Charles Louis de Secondat, Baron de Brede et de. *De l'esprit des laix. . . .* Geneva, 1748.

Morse, Jedidiah. *The History of America, in Two Books.* Philadelphia, [n.d.].

Nietzsche, Friedrich. *Von Nutzen und Nachteil der Historie für das Leben.* Vol. 1 of *Friedrich Nietzsche: Sämtliche Werke.* Ed. Giorgio Colli and Mazzino Montinari. 15 vols. Berlin: Walter de Gruyter, 1967–1977.

————. *On the Advantage and Disadvantage of History for Life.* Trans. Peter Preuss. Indianapolis: Hackett, 1980.

Noyes, Nicholas. *New-Englands Duty. . . .* Boston, 1698.

Oppenheimer, Robert. "Knowledge and the Structure of Culture." Helen Kenyon Lecture. Vassar College, 29 October 1958. Unpublished typescript. University of California, Los Angeles.

Orosius. See Paulus Orosius.

Otto of Freising. *Ottonis episcopi Frisingensis chronica sive historia de duabus civitatibus.* Ed. Adolf Hofmeister. Scriptores rerum Germanicarum in usum scholarum ex monumentis Germaniae historicis 45. Hannover: Bibliopoli Hahniani, 1912.

―――. *The Two Cities: A Chronicle of Universal History to the Year 1146 A.D.* Trans. Charles Christopher Mierow. Records of Civilization: Sources and Studies. New York: Columbia University Press, 1928.

Paris, Matthew. *Matthaei Parisiensis, monachi Sancti Albani, chronica majora.* Ed. Henry Richards Luard. 7 vols. Chronicles and Memorials of Great Britain and Ireland during the Middle Ages (Rolls Series) 54. London, 1877.

―――. *Matthew Paris's English History.* Trans. J. A. Giles. 3 vols. London, 1852–1854.

Paulus Orosius. *Pauli Orosii historiarum adversum paganos libri VII.* Ed. Carl Zangemeister. Leipzig, 1889.

―――. *Seven Books of History against the Pagans: The Apology of Paulus Orosius.* Trans. Irving Woodworth Raymond. Records of Civilization: Sources and Studies. New York: Columbia University Press, 1936.

Petrarch (Francesco Petrarca). *De sui ipsius et multorum ignorantia.* Ed. L. M. Capelli. Paris: Champion, 1906.

Ramsey, David. *The History of the American Revolution.* Philadelphia, 1789.

Robert de Boron. *Le Roman de l'estoire dou Graal.* Ed. William A. Nitze. Paris: Champion, 1927.

Santayana, George. *The Life of Reason: Or the Phases of Human Progress.* 5 vols. New York: Scribner's, 1905-1906.

Scaliger, Joseph Juste. *De emendatione temporum. . . .* Frankfurt, 1593.

Scottow, Joshua. *A Narrative of the Planting of the Massachusets Colony Anno 1628. With the Lords Signal Presence the First Thirty Years. Also a Caution from New-Englands Apostle, the Great Cotton, How to Escape the Calamity, which Might Befall Them or Their Posterity. And Confirmed by the Evangelist Norton with Prognosticks from the Famous Dr. Owen. concerning the Fate of these Churches, and Animadversions upon the Anger of God, in Sending of Evil Angels among Us. Published by Old Planters, the Authors of the Old Mens Tears.* Collections of the Massachusetts Historical Society 4th ser. 4. Boston, 1858.

Sewall, Samuel. *Phaenomena quaedam Apocalyptica ad Aspectum Novi Orbis Configurata: Or, Some Few Lines towards a Description of the New Heaven as It Makes to Those Who Stand upon the New Earth.* Boston, 1697.

Tertullian. *Apologeticus.* In *Tertullian: Apology, De spectaculis; Minucius Felix,* trans. T. R. Glover and Gerald H. Rendall. Loeb Classical Library. London: Heinemann; New York: Putnam's, 1931.

Thoreau, Henry David. *Henry David Thoreau: A Week on the Concord and Merrimack Rivers, Walden; or, Life in the Woods, The Maine Woods, Cape Cod.* Ed. Robert F. Sayre. New York: Library of America, 1985.

Ussher, James. *Jacobi Usserii Armachani annales veteris et Novi Testamenti. . . .* Paris, 1673.

Valla, Lorenzo. *The Treatise of Lorenzo Valla on the Donation of Constantine: Text and Translation into English.* Ed. and trans. Christopher B. Coleman. New Haven: Yale University Press; London: Oxford University Press, 1922.

Vasari, Giorgio. *Vite degli artefici.* In *Opere di Giorgio Vasari, pittore.* Biblioteca enciclopedica Italiana 2. Milan, 1829.

————. *Lives of the Artists.* Trans. George Bull. Harmondsworth: Penguin, 1965.

Vico, Giambattista. *Principi di scienza nuova di Giambattista Vico d'intorno alla comune natura delle nazioni.* 3rd ed. Naples, 1744.

————. *The New Science of Giambattista Vico: Unabridged Translation of the Third Edition (1744) with the Addition of "Practic of the New Science."* Trans. Thomas Goddard Bergin and Max Harold Fisch. Ithaca, N.Y.: Cornell University Press, 1984.

Webster, Noah. *History of the United States; To Which Is Prefixed a Brief Historical Account of Our English Ancestors, from the Dispersion of Babel, to Their Migration to America; and of the Conquest of South America, by the Spaniards.* New Haven, 1832.

Weems, Mason Locke. *The Life of George Washington; with Curious Anecdotes, Equally Honorable to Himself, and Exemplary to His Young Countrymen.* 6th ed. Philadelphia, 1808.

Weil, Simone. "A propos de la question coloniale, dans ses rapports avec le peuple français." In *Écrits historique et politiques.* Paris: Gallimard, 1960.

————. "East and West: Thoughts of the Colonial Problem." In *Selected Essays,* trans. Richard Rees. London: Oxford University Press, 1962.

William of Tyre. *Historia de rebus gestis in partibus transmarinis.* In *Guillaume de Tyr et ses continuateurs,* ed. Paulin Paris. 2 vols. Paris, 1879.

————. *A History of Deeds Done beyond the Sea.* Trans. Emily Atwater Babcock and A. C. Krey. 2 vols. Records of Civilization: Sources and Studies. New York: Columbia University Press, 1943.

Winthrop, John. *The History of New England from 1630 to 1649* [Journal]. Ed. James Savage. New ed. 2 vols. Boston, 1853.

Wittgenstein, Ludwig. *Culture and Value.* Ed. G. H. von Wright and Heikki Nyman. Trans. Peter Winch. Chicago: University of Chicago Press, 1984.

Wyclif, John. *Iohannis Wyclif: Tractatus de apostasia.* Ed. Michael Henry Dziewicki. London, 1888.

———. *Iohannis Wyclif: Tractatus de blasphemia.* Ed. Michael Henry Dziewicki. London: 1893.

———. *Iohannis Wyclif: Tractatus de ecclesia.* Ed. Johann Loserth. London: 1886.

———. *Johannis Wyclif: Tractatus de potestate pape.* Ed. Johann Loserth. London: Wyclif Society-Trübner, 1907.

Secondary Sources

General Works

Barnes, Harry Elmer. *A History of Historical Writing.* 2nd rev. ed. New York: Dover, 1962.

Brandon, S. G. F. *Time and Mankind: An Historical and Philosophical Study of Mankind's Attitude to the Phenomena of Change.* London: Hutchinson, 1951.

Braudel, Fernand. *Écrits sur l'histoire.* Paris: Flammarion, 1969.

———. *On History.* Trans. Sarah Matthews. Chicago: University of Chicago Press, 1980.

Breisach, Ernst. *Historiography: Ancient, Medieval, and Modern.* Chicago: University of Chicago Press, 1983.

Bury, J. B. *The Idea of Progress: An Inquiry into Its Origin and Growth.* New York: Macmillan, 1932.

Case, Shirley Jackson. *The Christian Philosophy of History.* Chicago: University of Chicago Press, 1943.

Certeau, Michel de. *L'Écriture de l'histoire.* Paris: Gallimard, 1975.

———. *The Writing of History.* Trans. Tom Conley. New York: Columbia University Press, 1988.

Collingwood, R. G. *The Idea of History.* 1946. London: Oxford University Press, 1956.

Commager, Henry Steele. *The Search for a Usable Past, and Other Essays in Historiography.* New York: Knopf, 1967.

Cook, Albert. *History/Writing.* Cambridge: Cambridge University Press, 1988.

Danto, Arthur C. *Narration and Knowledge.* New York: Columbia University Press, 1985.

Fischer, D. H. *Historians' Fallacies: Toward a Logic of Historical Thought.* New York: Harper, 1970.

Foucault, Michel. *L'Archéologie du savoir.* Paris: Gallimard, 1969.

———. *The Archaeology of Knowledge and the Discourse on Language.* Trans. A. M. Sheridan Smith. New York: Pantheon, 1982.

Gallie, W. B. *Philosophy and the Historical Understanding.* 2nd ed. New York: Schocken, 1968.

Gardiner, P., ed. *Theories of History: Readings from Classical and Contemporary Sources.* Free Press Textbooks in Philosophy. Glencoe, Ill.: Free Press, 1959.

Goldstein, Leon J. *Historical Knowing.* Austin: University of Texas Press, 1976.

Gorman, J. L. *The Expression of Historical Knowledge.* Edinburgh: Edinburgh University Press, 1982.

Hay, Denys. *Annalists and Historians: Western Historiography from the Eighth to the Eighteenth Centuries.* London: Methuen, 1977.

Himmelfarb, Gertrude. *The New History and the Old: Critical Essays and Reappraisals.* Cambridge, Mass.: Belknap Press, Harvard University Press, 1987.

Hook, Sidney, ed. *Philosophy and History: A Symposium.* New York: New York University Press, 1963.

Kuhn, Thomas. *The Structure of Scientific Revolutions.* Chicago: University of Chicago Press, 1959.

Lowenthal, David. *The Past Is a Foreign Country.* Cambridge: Cambridge University Press, 1985.

Löwith, Karl. *Meaning in History: The Theological Implications of the Philosophy of History.* Chicago: University of Chicago Press, 1949.

Mandelbaum, Maurice. *The Anatomy of Historical Knowledge.* Baltimore: Johns Hopkins University Press, 1977.

————. *The Problem of Historical Knowledge: An Answer to Relativism.* New York: Liveright, 1938.

Manuel, Frank E. *Shapes of Philosophical History.* Harry Camp Lectures at Stanford University. Stanford: Stanford University Press, 1965.

Marrou, Henri. *De la connaisance historique.* 5th ed. Paris: Seuil, 1966.

————. *The Meaning of History.* Trans. Robert J. Olsen. Baltimore: Helicon, 1966.

————. *Théologie de l'histoire.* Paris: Seuil, 1968.

————. *Time and Timeliness.* Trans. Violet Neville. New York: Sheed and Ward, 1969.

Meiland, Jack W. *Scepticism and Historical Knowledge.* New York: Random, 1965.

Mises, Ludwig von. *Theory and History: An Interpretation of Social and Economic Evolution.* New Haven: Yale University Press, 1957.

Munz, Peter. *The Shapes of Time: A New Look at the Philosophy of History.* Middletown, Conn.: Wesleyan University Press, 1977.

Nadel, George H., ed. *Studies in the Philosophy of History: Selected Essays from* History and Theory. New York: Harper, 1960.

————. "Philosophy of History before Historicism." *History and Theory: Studies in the Philosophy of History* 3 (1964): 219–315.

Nisbet, Robert A. *Social Change and History: Aspects of the Western Theory of Development.* New York: Oxford University Press, 1969.

Nota, John H. *Phenomenology and History.* Trans. Louis Grooten and John H. Nota. Chicago: Loyola University Press, 1967.

Patrides, C. A. *The Grand Design of God: The Literary Form of the Christian View of History.* Ideas and Forms in English Literature. London: Routledge; Toronto: University of Toronto Press, 1972.

Plumb, J. H. *The Death of the Past.* Boston: Houghton, 1970.

Popper, Karl R. *The Poverty of Historicism.* New York: Harper, 1964.

Rickert, Heinrich. *Die Grenzen der naturwissenschaftlichen Begriffsbildung: Eine logische Einleitung in die historischen Wissenschaften.* 5th ed. Tübingen: Mohr, 1929.

————. *The Limits of Concept Formation in Natural Science: A Logical Introduction to the Historical Sciences.* Ed. and trans. Guy Oakes. Texts in German Philosophy. Cambridge: Cambridge University Press, 1986.

Ricoeur, Paul. *Histoire et vérité.* 2nd ed. Paris: Seuil, 1964.

————. *History and Truth: Essays.* Trans. Charles A. Kelbley. Northwestern University Studies in Phenomenology and Existential Philosophy. Evanston, Ill.: Northwestern University Press, 1966.

————. *The Reality of the Historical Past.* Aquinas Lecture, 1984. Milwaukee: Marquette University Press, 1984.

————. *Temps et récit.* 3 vols. Paris: Seuil, 1983–1985.

————. *Time and Narrative.* Trans. Kathleen McLaughlin and David Pellaur. 3 vols. Chicago: University of Chicago Press, 1984–1988.

Shinn, Roger Lincoln. *Christianity and the Problem of History.* New York: Scribner's, 1953.

Simmel, Georg. *Die Probleme der Geschichtsphilosophie: Eine erkenntnistheoretische Studie.* Leipzig, 1892.

————. *The Problems of the Philosophy of History: An Epistemological Essay.* Trans. and ed. Guy Oakes. 2nd ed. New York: Free Press–Macmillan, 1977.

Stock, Brian. *Listening for the Text: On the Uses of the Past.* Baltimore: Johns Hopkins University Press, 1989.

Thompson, James Westfall. *A History of Historical Writing.* 2 vols. New York: Macmillan, 1942.

Tuveson, Ernest Lee. *Millennium and Utopia: A Study in the Background of the Idea of Progress.* New York: Harper, 1964.

White, Hayden. *The Content of the Form: Narrative Discourse and Historical Representation.* Baltimore: Johns Hopkins University Press, 1987.

————. *Tropics of Discourse: Essays in Cultural Criticism.* Baltimore: Johns Hopkins University Press, 1978.

Wilcox, Donald J. *The Measure of Times Past: Pre-Newtonian Chronologies and the Rhetoric of Relative Time.* Berkeley: University of California Press, 1987.

Early Christian and Medieval

Brandt, William J. *The Shape of Medieval History: Studies in Modes of Perception.* New Haven: Yale University Press, 1966.

Brown, Peter. *Augustine of Hippo: A Biography.* London: Faber, 1967.

————. *The World of Late Antiquity, AD 150–750.* History of European Civilization Library. New York: Harcourt, 1971.

Cullman, Oscar. *Christus und die Zeit; die urchliche Zeitund Geschichtsauffassung.* 3rd ed. Zürich: Evz-Verlag, 1962.

————. *Christ and Time: The Primitive Christian Conception of Time and History.* Trans. Floyd V. Wilson. London: SCM, 1962.

————. *Heil als Geschichte; heilsgeschichtliche Existenz im Neuen Testament.* Tübingen: Mohr, 1965.

Davis, R. H. C., and J. M. Wallace-Hadrill, eds. *The Writing of History in the Middle Ages: Essays Presented to Richard William Southern.* Oxford: Clarendon, 1981.

Dempf, Alois. *Sacrum Imperium: Geschichts- und Staatsphilosophie des Mittelalters und der politischen Renaissance.* 2nd ed. Darmstadt: Wissenschaftliche Buchgemeinschaft, 1954.

Emmerson, Richard Kenneth. *Antichrist in the Middle Ages: A Study of Medieval Apocalypticism, Art, and Literature.* Seattle: University of Washington Press, 1981.

Freund, Walter. *Modernus und andere Zeitbegriffe des Mittelalters.* Cologne: Böhlau, 1957.

Grant, Robert M. *Eusebius as Church Historian.* Oxford: Clarendon, 1980.

Guenée, Bernard. "Histoires, Annales, Chroniques." *Annales: Economies, Sociétés, Civilisations* 28 (1972): 997–1016.

————. *Histoire et culture historique dans L'Occident médiéval.* Paris: Aubier Montaigne, 1980.

Hanning, Robert W. *The Vision of History in Early Britain: From Gildas to Geoffrey of Monmouth.* New York: Columbia University Press, 1966.

Keyes, Gordon L. *Christian Faith and the Interpretation of History: A Study of St. Augustine's Philosophy of History.* Lincoln: University of Nebraska Press, 1966.

Lacroix, Benoit. *Orose et ses idées.* Publications de l'Institute d'études médiévales 18. Montreal: Institute d'études médiévales, 1965.

Lewis, C. S. *The Discarded Image: An Introduction to Medieval and Renaissance Literature.* Cambridge: Cambridge University Press, 1964.

Loewenich, Walther von. *Augustin und das christliche Geschichtsdenken.* Munich: Kaiser, 1947.

Maier, Franz Georg. *Augustin und das antike Rom: Untersuchungen zur Auseinandersetzung Augustins mit römischer Geschichte, Romgedanken und politischer Religiosität.* Tübinger Beiträger zur Altertumswissenschaft 39. Stuttgart: Kohlhammer, 1955.

Markus, R. A. *Saeculum: History and Society in the Theology of St. Augustin.* Cambridge: Cambridge University Press, 1970.

Marrou, Henri. *Saint Augustin et la fin de la culture antique.* Paris: Boccard, 1958.

―――. *Saint Augustine et l'augustinisme.* Paris: Seuil, 1958.

―――. *St. Augustine and His Influence through the Ages.* Trans. Patrick Hepburne-Scott and Edmund Hill. New York: Harper, 1957.

Mierow, Charles Christopher. "Otto of Freising and His Two Cities Theory." *Philological Quarterly* 24 (1945): 97–105.

Milburn, Robert L. P. *Early Christian Interpretations of History.* London: Black; New York: Harper, 1954.

Mommsen, Theodore E. "Augustine and the Christian Idea of Progress: The Background of the City of God." *Journal of the History of Ideas* 12 (1951): 346–74.

Mosshammer, Alden A. *The Chronicle of Eusebius and Greek Chronographic Tradition.* Lewisburg, Penn.: Bucknell University Press, 1979.

Patterson, L. G. *God and History in Early Christian Thought.* New York: Seabury, 1967.

Petry, R. C. "Three Medieval Chroniclers: Monastic Historiography and Biblical Eschatology in Hugh of St. Victor, Otto of Freising, and Ordericus Vitalis." *Church History* 34 (1965): 282–93.

Pickering, F. P. *Augustinus oder Boethius?—Geschichtsschreibung und epische Dichtung im Mittelalter und in der Neuzeit.* Philologische Studien und Quellen 39. Berlin: Schmidt, 1967.

Poole, Reginald. *Chronicles and Annals: A Brief Outline of Their Origin and Growth.* Oxford: Clarendon, 1926.

Reeves, M. *The Influence of Prophecy in the Later Middle Ages: A Study in Joachimism.* Oxford: Clarendon, 1969.

Ruotolo, Giuseppe. *La filosofia della storia e la citta di Dio.* 2nd ed. Rome: Zuffi, 1950.

Smalley, Beryl. *Historians in the Middle Ages.* New York: Scribner's, 1974.

Spörl, J. "Das Alte und das Neue im Mittelalter: Studien zum Problem des mittelalterlichen Forschrittsbewusstseins." *Historisches Jahrbuch* 50 (1930): 297–341 and 498–524.

Sterns, Indrikis. *The Greater Medieval Historians: An Interpretation and Bibliography.* Lanham, Md.: University Press of America, 1980.

Stevenson, James. *Studies in Eusebius.* Cambridge: Cambridge University Press, 1929.

Varga, Lucie. *Das Schlagwort vom "finsteren Mittelalter."* Baden: Rohrer, 1932.

Wallace-Hadrill, D. S. *Eusebius of Caesaria.* London: Mowbray, 1960.

Reformation and Renaissance

Baker, Herschel. *The Race of Time: Three Lectures on Renaissance Historiography.* Toronto: University of Toronto Press, 1967.

Bauckham, Richard. *Tudor Apocalypse: Sixteenth Century Apocalypticism, Millenarianism and the English Reformation: From John Bale to John Foxe and Thomas Brightman.* Courtney Library of Reformation Classics 8. Oxford: Sutton Courtney, [1978?].

Bayley, C. C. "Petrarch, Charles IV, and the 'renovatio imperii.'" *Speculum: A Journal of Medieval Studies* 17 (1942): 323–41.

Cochrane, Eric W. *Historians and Historiography in the Italian Renaissance.* Chicago: University of Chicago Press, 1981.

Dean, Leonard Fellows. *Tudor Theories of Historical Writing.* University of Michigan Contributions in Modern Philology 1. Ann Arbor: University of Michigan Press, 1947.

Eisenstein, Elizabeth L. *The Printing Press as an Agent of Change: Communications and Cultural Transformations in Early-Modern Europe.* 2 vols. Cambridge: Cambridge University Press, 1979.

Fairfield, Leslie P. *John Bale: Mythmaker for the English Reformation.* West Lafayette, Ind.: Purdue University Press, 1976.

Ferguson, Arthur B. *Clio Unbound: Perception of the Social and Cultural Past in Renaissance England.* Duke Monographs in Medieval and Renaissance Studies 2. Durham, N.C.: Duke University Press, 1979.

Ferguson, W. K. *Europe in Transition: 1300–1520.* Boston: Houghton, 1962.

Fryde, E. B. *Humanism and Renaissance Historiography.* London: Hambledon, 1983.

Fussner, F. Smith. *The Historical Revolution: English Historical Writing and Thought, 1580–1640.* Westport, Conn.: Greenwood, 1962.

Goetz, Werner. *Translatio imperii; ein Beitrag zur Geschichte des Geschichtsdenkens und der politischen Theorien in Mittelalter und in der frühen Neuzeit.* Tübingen: Mohr, 1958.

Green, Louis. *Chronicle into History: An Essay on the Interpretation of History in Florentine Fourteenth-Century Chronicles.* Cambridge: Cambridge University Press, 1972.

Guibbory, Achsah. "Francis Bacon's View of History: The Cycles of Error and the Progress of Truth." *Journal of English and Germanic Philology* 74 (1975): 336–50.

Haller, William. *The Elect Nation: The Meaning and Relevance of Foxe's "Book of Martyrs."* New York: Harper, 1963.

————. *The Rise of Puritanism: Or, the Way to the New Jerusalem as Set Forth in Pulpit and Press from Thomas Cartwright to John Lilburn and John Milton, 1570–1643.* New York: Columbia University Press, 1938.

Hay, Denys. *Flavio Biondo and the Middle Ages: Italian Lecture, British Academy, 1959.* Proceedings of the British Academy 45: 98–127. London: Oxford University Press, 1959.

Headley, John M. *Luther's View of Church History.* Yale Publications in Religion 6. New Haven: Yale University Press, 1963.

Kuhn, Thomas. *The Copernican Revolution: Planetary Astronomy in the Development of Western Thought.* Cambridge, Mass.: Harvard University Press, 1957.

Levy, F. J. *Tudor Historical Thought.* San Marino, Calif.: Huntingdon Library, 1967.

Meiland, Jack W. *Scepticism and Historical Knowledge.* Random House Studies in Philosophy 5. New York: Random, 1965.

Mommsen, Theodore E. "Petrarch's Conception of the 'Dark Ages.'" *Speculum: A Journal of Medieval Studies* 17 (1942): 226–42.

Phillips, M. *Francesco Guicciardini: The Historian's Craft.* University of Toronto Romance Series. Toronto: University of Toronto Press, 1977.

————. "Machiavelli, Guicciardini and the Tradition of Vernacular Historiography in Florence." *American Historical Review* 84 (1979): 86–105.

Pullapilly, Cyriac K. *Caesar Baronius, Counter-Reformation Historian.* Notre Dame, Ind.: University of Notre Dame Press, 1975.

Quinones, Ricardo J. *The Renaissance Discovery of Time.* Harvard Studies in Comparative Literature 31. Cambridge, Mass.: Harvard University Press, 1972.

Stinger, Charles L. *Humanism and the Church Fathers: Ambrogio Traversari (1386–1439) and Christian Antiquity in the Italian Renaissance.* Albany: State University of New York Press, 1977.

Struever, Nancy S. *The Language of History in the Renaissance: Rhetoric and Historical Consciousness in Florentine Humanism.* Princeton: Princeton University Press, 1970.

Ullman, B. "Leonardo Bruni and Humanistic Historiography." *Medievalia et Humanistica* 4 (1946): 45–61.

Wilcox, Donald J. *The Development of Florentine Humanist Historiography in the Fifteenth Century.* Harvard Historical Studies 82. Cambridge, Mass.: Harvard University Press, 1969.

————. *In Search of God and Self: Renaissance and Reformation Thought.* Boston: Houghton, 1975.

Zakai, Avihu. "Reformation, History, and Eschatology in English Protestantism." *History and Theory: Studies in the Philosophy of History* 26 (1987): 300–318.

Colonial American

Bercovitch, Sacvan. *The American Jeremiad.* Madison: University of Wisconsin Press, 1978.

————. "The Historiography of Johnson's *Wonder-Working Providence.*" *Essex Institute Historical Collections* 104 (1968): 138–61.

————. "New England Epic: Cotton Mather's *Magnalia Christi Americana.*" *ELH* 33 (1966): 337–50.

————. *The Puritan Origins of the American Self.* New Haven: Yale University Press, 1975.

Bercovitch, Sacvan, ed. *The American Puritan Imagination: Essays in Revaluation.* London: Cambridge University Press, 1974.

————, ed. *Typology and Early American Literature.* [Amherst]: University of Massachusetts Press, 1972.

Brumm, Ursula. "Edward Johnson's *Wonder-Working Providence* and the Puritan Conception of History." *Jahrbuch für Amerikastudien* 14 (1969): 140–51.

————. *Die religiöse Typologie im amerikanischen Denken; ihre Bedeutung für die amerikanische Literatur- und Geistesgeschichte.* Studien zur amerikanische Literatur und Geschichte 2. Leiden: Brill, 1963.

Daly, Robert. "William Bradford's Vision of History." *American Literature* 44 (1973): 557–69.

Delbanco, Andrew. "The Puritan Errand Re-Viewed." *Journal of American Studies* 18 (1984): 343–60.

————. *The Puritan Ordeal.* Cambridge, Mass.: Harvard University Press, 1989.

Gallagher, Edward J. "An Overview of Edward Johnson's *Wonder-Working Providence.*" *Early American Literature* 5 (1971): 30–49.

Gay, Peter. *A Loss of Mastery: Puritan Historians in Colonial America.* Berkeley: University of California Press, 1966.

Griffith, John. "*Of Plymouth Plantation* as a Mercantile Epic." *Arizona Quarterly* 28 (1972): 231–42.

Heimert, Alan. *Religion and the American Mind: From the Great Awakening to the Revolution.* Cambridge, Mass.: Harvard University Press, 1966.

Heimert, Alan, and Andrew Delbanco, eds. *The Puritans in America: A Narrative Anthology.* Cambridge, Mass.: Harvard University Press, 1985.

Hill, Christopher. *Puritanism and Revolution.* New York: Schocken, 1964.

Hovey, Kenneth Alan. "The Theology of History in *Of Plymouth Plantation* and Its Predecessors." *Early American Literature* 10 (1975): 47–66.

Howard, Alan. "Art and History in Bradford's *Of Plymouth Plantation.*" *William and Mary Quarterly* 3rd ser. 28 (1971): 237–66.

Jennings, Francis. *The Invasion of America: Indians, Colonialism and the Cant of Conquest.* Chapel Hill: University of North Carolina Press, 1975.

Kibbey, Ann. *The Interpretation of Material Shapes in Puritanism: A Study of Rhetoric, Prejudice, and Violence.* Cambridge Studies in American Literature and Culture. Cambridge: Cambridge University Press, 1986.

Levin, David. *Cotton Mather: The Young Life of the Lord's Remembrancer, 1663–1703.* Cambridge, Mass.: Harvard University Press, 1978.

———. "William Bradford: The Value of Puritan Historiography." In *Major Writers of Early American Literature*, ed. Everett Emerson. Madison: University of Wisconsin Press, 1972.

Lowance, Mason I. *The Language of Canaan: Metaphor and Symbol in New England from the Puritans to the Transcendentalists.* Cambridge, Mass.: Harvard University Press, 1980.

———. "Typology and the New England Way: Cotton Mather and the Exegesis of Biblical Types." *Early American Literature* 4 (1969): 15–37.

Miller, Perry. *Errand into the Wilderness.* Cambridge, Mass.: Harvard University Press, 1958.

———. *The New England Mind: From Colony to Province.* Cambridge, Mass.: Belknap Press, Harvard University Press, 1953.

———. *The New England Mind: The Seventeenth Century.* New York: Harper, 1939

Miller, Perry, and Thomas H. Johnson. *The Puritans: A Sourcebook of Their Writings.* 2 vols. New York: Harper, 1963.

Murdock, Kenneth B. "Clio in the Wilderness: History and Biography in Puritan New England." *Church History* 24 (1955): 221–38. Rpt. in *Early American Literature* 6 (1972): 201–19.

Powers, Dennis. "Purpose and Design in Joshua Scottow's *Narrative.*" *Early American Literature* 18 (1984): 275–90.

Rosenmeier, Jesper. "'With my owne eyes': William Bradford's *Of Plymouth Plantation.*" In *The American Puritan Imagination: Essays in Revaluation*, ed. Sacvan Bercovitch. London: Cambridge University Press, 1974. Also in *Typology and Early American Literature*, ed. Sacvan Bercovitch. [Amherst]: University of Massachusetts Press, 1972.

Silverman, Kenneth. *The Life and Times of Cotton Mather.* New York: Harper, 1984.

Tichi, Cecilia. *New World, New Earth: Environmental Reform in American*

Literature from the Puritans through Whitman. New Haven: Yale University Press, 1979.

————. "The Puritan Historians and Their New Jerusalem." *Early American Literature* 6 (1971): 143–55.

Tuveson, Ernest Lee. *Redeemer Nation: The Idea of America's Millennial Role*. Chicago: University of Chicago Press, 1968.

Vaughan, Alden T. *New England Frontier: Puritans and Indians, 1620–1675*. Boston: Little, 1965.

Walsh, James P. "Holy Time and Sacred Space in Puritan New England." *American Quarterly* 32 (1980): 79–95.

Wenska, Walter P. "Bradford's Two Histories: Pattern and Paradigm in *Of Plymouth Plantation*." *Early American Literature* 13 (1978): 151–64.

Ziff, Larzer. *Puritanism in America: New Culture in a New World*. New York: Viking, 1973.

Enlightenment to Modern

Barzun, Jacques. *Race: A Study in Superstition*. New York: Harcourt, 1965.

Berlin, Isaiah. *The Age of Enlightenment: The Eighteenth-Century Philosophers*. Boston: Houghton, 1956.

————. *Vico and Herder: Two Studies in the History of Ideas*. New York: Viking, 1976.

Berman, Marshall. *All That Is Solid Melts into Air: The Experience of Modernity*. Harmondsworth: Penguin, 1988.

Bowler, Peter J. *Evolution: The History of an Idea*. Berkeley: University of California Press, 1984.

Bury, J. B. *The Idea of Progress: An Inquiry into Its Origin and Growth*. London: Macmillan, 1920.

Cassirer, Ernst. *Die Philosophie der Aufklärung*. Grundriss der philosophischen Wissenschaften. Tübingen: Mohr, 1932.

————. *The Philosophy of the Enlightenment*. Trans. Fritz C. A. Koelin and James P. Pettegrove. Princeton: Princeton University Press, 1951.

Clive, John. *Macaulay: The Shaping of the Historian*. New York: Knopf, 1973.

Cooke, Jacob E. *Frederick Bancroft, Historian*. Norman: University of Oklahoma Press, 1957.

Cragg, Gerald Robertson. *From Puritanism to the Age of Reason: A Study of Changes in Religious Thought within the Church of England, 1660–1700*. Cambridge: Cambridge University Press, 1950.

Crow, Charles. "The Emergence of Progressive History." *Journal of the History of Ideas* 27 (1966): 109–24.

Donovan, T. *Henry Adams and Brooks Adams: The Education of Two American Historians*. Norman: University of Oklahoma Press, 1961.

Dray, William H., ed. *Philosophical Analysis and History.* Sources in Contemporary Philosophy. New York: Harper, 1966.

Fueter, E. *Geschichte der neueren Historiographie.* 3rd ed. Munich: Oldenbourg, 1936.

Gay, Peter. *The Enlightenment, An Interpretation: The Rise of Modern Paganism.* New York: Knopf, 1966.

Haller, John S. *Outcasts from Evolution: Scientific Attitudes of Racial Inferiority, 1859–1900.* Urbana: University of Illinois Press, 1975.

Hamill, Paul J. "Science as Ideology: The Case of the Amateur, Henry Adams." *Canadian Review of American Studies* 12 (1981): 21–35.

Hofstadter, Richard. *The Progressive Historians: Turner, Beard, Parrington.* New York: Knopf, 1970.

Jordy, William H. *Henry Adams: Scientific Historian.* Yale Historical Publications Studies 16. New Haven: Yale University Press, 1952.

Koch, Gustav Adolf. *Republican Religion: The American Revolution and the Cult of Reason.* New York: Holt, 1933. Republished as *Religion of the American Enlightenment.* New York: Crowell, 1968.

Koselleck, Reinhart. *Kritik und Krise: Eine Studie zur Pathogenese der bürgerlichen Welt.* Freiburg: Karl Alber, 1959.

———. *Critique and Crisis: Enlightenment and the Pathogenesis of Modern Society.* Cambridge, Mass.: MIT Press, 1988.

Kronick, Joseph G. "The Limits of Contradiction: Irony and History in Hegel and Henry Adams." *CLIO: A Journal of Literature, History, and the Philosophy of History* 15 (1986): 391–410.

Lesser, Wayne. "Criticism, Literary History and the Paradigm: *The Education of Henry Adams*." *PMLA: Publications of the Modern Language Association of America* 97 (1982): 378–94.

Liebel, H. P. "The Enlightenment and the Rise of Historicism in German Thought." *Eighteenth-Century Studies* 4 (1970): 359–85.

Lienesch, Michael. *New Order of the Ages: Time, the Constitution, and the Making of Modern American Political Thought.* Princeton: Princeton University Press, 1988.

Manuel, Frank Edward. *The Eighteenth Century Confronts the Gods.* Cambridge, Mass.: Harvard University Press, 1959.

———. *Isaac Newton, Historian.* Cambridge, Mass.: Harvard University Press, 1963.

Melton, J. Gordon. "Spiritualization and Reaffirmation: What Really Happens When Prophecy Fails." *American Studies* 26 (1985): 12–29.

Mink, Louis O. "Philosophical Analysis and Historical Understanding." *Review of Metaphysics* 21 (1968): 667–98.

Montagu, Ashley. *Man's Most Dangerous Myth: The Fallacy of Race.* 5th ed. Oxford: Oxford University Press, 1974.

————. *Race, Society and Humanity*. New York: Van Nostrand Reinhold, 1963.

Munford, Howard N. "Henry Adams: The Limitations of Science." In *Critical Essays on Henry Adams*, ed. Earl N. Harbert. Boston: Hall, 1981.

Murray, Michael. *Modern Philosophy of History: Its Origin and Destination*. The Hague: Nijhoff, 1970.

Pagliaro, Harold E, ed. *Racism in the Eighteenth Century*. Studies in Eighteenth-Century Literature 3. Cleveland: Case Western Reserve University Press, 1973.

Quinones, Ricardo J. *Mapping Literary Modernism: Time and Development*. Princeton: Princeton University Press, 1985.

Reill, Peter Hans. *The German Enlightenment and the Rise of Historicism*. Berkeley: University of California Press, 1975.

Sallee, Jarel C. "Henry Adams' Emersonian Education." *ESQ: A Journal of the American Renaissance* 27 (1981): 38–46.

Snyder, Louis L. *The Idea of Racialism: Its Meaning and History*. Princeton: Princeton University Press, 1962.

Stromberg, Roland N. *Religious Liberalism in Eighteenth-Century England*. London: Oxford University Press, 1954.

Trevor-Roper, Hugh. "The Historical Philosophy of the Enlightenment." *Studies on Voltaire and the Eighteenth Century* 27 (1963): 1667–87.

White, Hayden. "The Burden of History." *History and Theory: Studies in the Philosophy of History* 5 (1966): 111–34.

————. "Foucault Decoded: Notes from Underground." *History and Theory: Studies in the Philosophy of History* 12 (1973): 23–34.

————. *Metahistory: The Historical Imagination in Nineteenth-Century Europe*. Baltimore: Johns Hopkins University Press, 1973.

Wilkins, B. T. *Hegel's Philosophy of History*. Ithaca, N.Y.: Cornell University Press, 1974.

Index

224